The Tropical Deciduous
Forest of Alamos

The Tropical Deciduous Forest of Alamos

Biodiversity of a Threatened Ecosystem in Mexico

EDITED BY

ROBERT H. ROBICHAUX

& DAVID A. YETMAN

THE UNIVERSITY OF ARIZONA PRESS

TUCSON

The University of Arizona Press
www.uapress.arizone.edu

© 2000 The Arizona Board of Regents
All rights reserved. Published 2000
First paperback edition 2016

Printed in the United States of America
21   20   19   18   17   16   7   6   5   4   3   2

ISBN-13: 978-0-8165-1922-4 (cloth)
ISBN-13: 978-0-8165-3416-6 (paper)

Cover illustration: *Pseudobombax palmeri*, by Anne E. Gondor

Library of Congress Cataloguing-in-Publication Data
The tropical deciduous forest of Alamos : biodiversity of a threatened ecosystem
in Mexico / edited by Robert H. Robichaux and David A. Yetman.
    p.   cm.
 Includes bibliographical references and index.
 ISBN 0-8165-1922-6 (cloth : alk. paper)
 1. Biological diversity—Mexico—Alamos (Sonora). I. Robichaux, Robert Hall,
1953–. II. Yetman, David, 1941–
 QH107.T76   2000
 333.95'11'097217—dc21
                 99-050867
                 CIP

British Library Cataloguing-in-Publication Data
A catalogue for this book is available from the British Library.

∞ This paper meets the requirements of ANSI/NISO Z39.48-1992 (Permanence
of Paper).

CONTENTS

PREFACE

Tropical deciduous forest, complete with fig and kapok trees, boa constrictors, macaws, and jungle cats, lies within an easy day's drive south of the U.S.–Mexican border. This may come as a surprise to many biologists and naturalists who are accustomed to the idea that tropical forests grow mainly within the latitudinal boundaries of the Tropics. Yet near the colonial town of Alamos in Sonora, Mexico, only a few hundred kilometers south of the border, tropical deciduous forest approaches its northern distributional limit in the Western Hemisphere (fig. P.1).

As highlighted in this volume, the tropical deciduous forest of the Alamos region contains a spectacular array of plant and animal life. Paul Martin and David Yetman (chapter 1) introduce this ecosystem, providing an overview of its biodiversity, the human uses of its rich resources, and the threats to its integrity. Angelina Martínez-Yrízar et al. (chapter 2) compare the structure and functioning of the tropical deciduous forest of the Alamos region with those of other areas in western Mexico. Thomas Van Devender et al. (chapter 3) analyze the vegetation and flora of the tropical deciduous forest, selecting as a focal point for analysis the Río Cuchujaqui, which flows from the mountains northeast of Alamos south into the Río Fuerte. The comprehensive plant list for the Río Cuchujaqui area well illustrates the plant diversity of the forest.

In chapter 4, David Yetman et al. examine the many tree species of the tropical deciduous forest that are used by the Mayo people of Sonora. Over the centuries Mayos and other indigenous peoples of the region have also developed numerous varieties of domesticated plants that flourish within the limits of this ecosystem. Barney Burns et al. (chapter 5) enumerate these domesticates and discuss their role in native cultures.

In addition to its rich flora, the tropical deciduous forest of the Alamos region contains a high diversity of native vertebrates. Cecil Schwalbe and Charles Lowe (chapter 6) examine the diversity and distributions of amphibians, reptiles, and mammals in the region. In chapter 7, Stephen Russell provides a comprehensive analysis of the birds of the Alamos area, including their ecological distributions and seasonal behaviors.

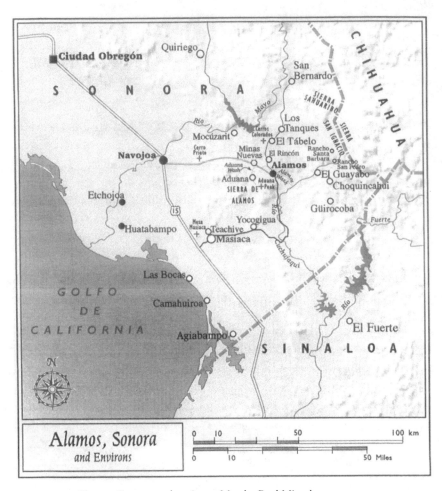

FIGURE P.I. Alamos, Sonora, and environs. Map by Paul Mirocha.

By providing a broad introduction to the biodiversity of the tropical deciduous forest near its northern distributional limit, this volume complements and extends the excellent recent volumes by Bullock et al. (1995) and Martin et al. (1998). Despite the recent surge of scientific interest in the Alamos region, much remains to be learned, especially about organisms such as insects, whose absence from this volume underscores the need for additional research.

This volume had its origins in a scientific gathering sponsored by the Arizona-Sonora Desert Museum in Tucson. The support of the museum staff is gratefully acknowledged, as is the help of many others in the United

States and Mexico whose lives are intertwined with the tropical deciduous forest of the Alamos region.

REFERENCES

Bullock, S. H., H. A. Mooney, and E. Medina, eds. 1995. *Seasonally dry tropical forests*. Cambridge: Cambridge University Press.

Martin, P. S., D. A. Yetman, M. E. Fishbein, P. D. Jenkins, T. R. Van Devender, and R. Wilson, eds. 1998. *Gentry's Río Mayo plants: The tropical deciduous forest and environs of northwest Mexico*. Tucson: University of Arizona Press.

The Tropical Deciduous Forest of Alamos

INTRODUCTION AND PROSPECT

Secrets of a Tropical Deciduous Forest

Paul S. Martin and David A. Yetman

The tropical deciduous forests of northwestern Mexico, along with the people who live there and know the forests best, stand at a turning point. The Río Mayo region of southern Sonora and western Chihuahua harbors more than 100 species of tropical trees in 36 families, 48 species of orchids, 6 species of parrots, mountain lions, jaguars, and according to local legend, a big cat known as the *onza*. Although destruction of tropical deciduous forest (TDF) and thorn forest in favor of the African exotic, buffelgrass, has accelerated, TDF endures near its northern limit, thanks to the extraordinary roughness of the foothills and *barrancas* (cliffs and gorges) of the Sierra Madre Occidental. Roughness provides a multitude of plant habitats, including the moist floors of slot canyons with their vertical walls that shelter tropical trees from the blazing sun of the dry season.

It may come as a surprise that indigenous Americans (Guarijío, Mayo, Yaqui, and Tarahumara) survive in this region. Some grow modest crops of locally adapted, genetically variable, native varieties of corn, beans, and other cultigens. They harvest forest products and raise livestock (see chapter 5). In the absence of wage labor they subsist on the land, which they know intimately, without benefit of a reservation, speaking their own tongue. This simple, self-reliant life exists, not in some remote tributary of the Upper Amazon in eastern Peru, but within 100 km of Ciudad Obregón, Sonora's agricultural metropolis of half a million people and an easy day's drive from the U.S.–Mexico border. The contrast could not be more extreme.

Plant geographers include TDF or monsoon forest within their sequence of world plant formations. Although one might not realize this from TV programs, rain forest is not the only important or the only threatened forest type in the tropics. Tropical forests form a graded series from wet to dry. In wet climates there are tall (> 30 m), unarmed, buttressed trees with medium-sized (> 8 cm long) or large, simple, evergreen leaves typically with drip tips.

In dry climates one finds short (< 5 m) trees commonly armed with thorns, lacking buttresses, and covered with small (< 8 cm) or compound leaves that lack drip tips.

TDF trees are in the middle of the gradient, taller than those of the thorn forest and shorter than those of the rain forest. Mostly deciduous, the leaves are larger than those found in the thorn forest and smaller than those of the rain forest. Some leaves have drip tips. Thorns or spines occur in some TDF species; for example, large pyramidal thorns cover the trunks of young kapok trees, locally known as *pochote* or *baogua* (*Ceiba acuminata*). Even without leaves, flowers, or fruit, many trees of the dry tropical forest can easily be recognized by the texture and color of their bark, such as palo santo, or by the architecture of their trunks and branches, such as brasil.

TDF goes by other names, including *monsoon forest* and *dry tropical forest*. In Spanish it is called the *selva baja caducifolia* or *bosque tropical caducifolio* (see chapter 2). In French it is known as *forêt tropicale caducifoliée* (Puig 1976). In his pioneering study of the northern tropics of the Río Mayo region, Gentry (1942) called it *short tree forest*. Local people around Alamos, Sonora, know it as the *monte mojino*, or "gray woods," from the color of many bare branches and trunks in the lengthy dry season (see chapter 4). In a popular book (Bowden et al. 1993), the authors appropriately named the monte around Alamos, Sonora, the *secret forest*. The forest type is poorly known, not only in Sonora but also elsewhere in Mexico where it occupies both coasts and in Central America, the West Indies, South America, and the tropics of other continents. In the Guanacaste of Costa Rica, TDF has been so thoroughly converted into pasture that ecologist Dan Janzen undertook to replant some with the cooperation of local cattle ranchers (Allen 1988).

What are four species of tropical figs (*Ficus*), kapok, palo barril (*Cochlospermum vitifolium*), and dozens of other tropical trees (see chapters 3 and 4), some with close relatives as far away as Paraguay, doing up here in Sonora and Chihuahua, north of 27°, at the latitude of Laredo, Texas? South Texas has tall palms and a few small tropical trees at the northern limit of their range, but nothing like this. West of the Sierra Madre Occidental, TDF ranges so far north it grows in foothills within 250 km (150 mi) of the Arizona border (31°20'N), well to the north of the tropic of Cancer (22°30'N) in Sinaloa. Apparently there is more to being tropical than the position of the noonday sun at the summer solstice.

Two climatic features are crucial: the Mexican monsoon and the outbursts of continental polar air masses known as *northers*. At the start of the summer rains, TDF trees rapidly switch from dry and leafless to wet and leafy, soaking up a July rainfall of 200 mm (8 in.) or more (Douglas et al. 1993). TDF hugs the deeply dissected skirts of the western edge of the Sierra Madre Occidental, which crests in mossy forests of pine and fir at elevations exceeding 2,700 m (9,000 ft). Summer rains are essential for tropical biota. To the north, as the rains weaken, tropical trees shrink in size; some become shrubs before they disappear entirely. No Arizona stations at any elevation average the meter or more of precipitation recorded in the pine-oak forests of southeastern Sonora and western Chihuahua.

The sierra shelters the tropical biota of western Mexico, eastern Sonora, Sinaloa, and western Chihuahua in another way. It wards off the incursions of El Norte, the "Blue Norther." In some winters these frigid arctic air masses sweep south out of Siberia and Canada to freeze cattle on open range in the Plains states, grapefruit in the lower Rio Grande valley in southern Texas, and tropical vegetation as far south as the Yucatán. Such freezes are very rare in the TDF of northwestern Mexico.

Thanks to reliable summer rain and protection from frost, the foothills and barrancas in northwestern Mexico shelter many tropical plants and animals. Wild avocados, wild figs, epiphytic orchids, tank bromeliads, and military macaws, to name a few, are remarkable to find this far north in North America. As shrubs a few species of tropical trees typical of TDF even manage to survive at elevations above 800 m in the mountains of southern Arizona. These include chilicote or coral bean (*Erythrina flabelliformis*), tepeguaje (*Lysiloma watsonii*), and yellow trumpet or palo de arco (*Tecoma stans*). When a rare killing frost like that of December 7, 1978, destroys their tops, they recover by sprouting from the base.

To get an up-close, hands-on look at representative TDF, we recommend a trip to Alamos, Sonora, a study area for authors of chapters in this volume. Alamos is a 5-hour drive south of Hermosillo, the capital of Sonora, and a 10-hour drive from Tucson, Arizona. In Sonora all but the last 50 km are on Highway 15, a four-lane, high-speed toll road. East of Navojoa the road to Alamos passes chicken and pig ranches. Charolais cattle graze in buffelgrass pastures punctuated by a few spreading trees, such as palo blanco (*Piscidia mollis*) and palo santo (*Ipomoea arborescens*), spared from the days when this was a tropical thorn forest. While they last (we doubt they can regenerate), the "keeper trees," those deliberately spared the bulldozer for their reproductive potential, show the height of the former forest. Closer to Alamos in wetter and rougher country, the road tunnels into a few miles of forested foothills. Here travelers can experience a touch of the TDF of the Sierra de Alamos.

With swimming, snorkeling, boating, golfing, time-sharing condominiums, and Club Med social life concentrated along the coast, relatively few tourists have found their way east of Highway 15 to Alamos. That may change if ecotourists discover that one does not have to fly to Belem in Brazil, Belize in Central America, or Borneo in Southeast Asia to experience tropical habitats. In addition, those enchanted by historical common-wall buildings with French doors and grillwork revealing shady interior garden patios will be delighted with Alamos, northern Mexico's most attractive and laid-back colonial mining town. Ecotourists will find the strange, exotic TDF in plain view on all sides of the *mirador* (lookout site) atop a hill on the eastern edge of Alamos.

To enter the forest and the newly designated reserve (Area de Protección de Flora y de Fauna) above Alamos, take the Magnolia Trail (Camino de Magnolia) a mile or two out of Parque Chalatón on the west side of town. Go with a local guide so you don't stumble into some private garden where you are not wanted. Or you can follow dirt roads to the Río Cuchujaqui either east or south of Alamos, the region with some 750 species of vascular plants, including dozens of species of tropical trees as inventoried in chapter 3.

There are many interesting trees along the route, such as brasil *(Haematoxylum)*, which is cut for indestructible stacked wood corrals; mauto *(Lysiloma divaricatum)*, the most common tree of the uncut forest and known to locals by three common names for three allegedly different trees; chírahui or boatthorn acacia *(Acacia cochliacantha)*, with flaring thorns on young trees; and four species of columnar cacti, including organpipe *(Stenocereus thurberi)*, soon to be sacrificed in its coastal heartland as the Huites Dam fills. Near the Río Cuchujaqui beyond Sabanito Sur at El Guayabo grow a few spreading cuajilote *(Pseudobombax palmeri)*. Their flowers appear in May and resemble an old-fashioned shaving brush. They open at sunset and attract pollinating long-nosed bats who dart in, take a pull at the nectar, receive a dab of pollen, and are gone, all in one second.

The "common" names found in this volume, such as *cuajilote*, may sound strange to ecologists, naturalists, or visitors from elsewhere. They are, nevertheless, the touchstone, the lingua franca for a beginner who finds himself in TDF in Sonora or northern Sinaloa. The names, like the plants, are native to the region. This knowledge is free for the asking.

What do we mean by that? The drill is simple: Just ask. Break off a branch of the tree you want identified and take it to a friendly *vaquero* or *paisano* (a local rancher; a passerby—whether Mayo, Guarijío, or mestizo—who lives in the region; or a woman who manages the rustic craft shop in Aduana). Ask politely, *"Dispenseme. ¿Cómo se llama este?"* ("Excuse me. What is this called?"). If your new contact does not know the name of the tree or shrub that you, the novice, found interesting, chances are he or she can find someone who does. The tables of names in chapters 3 and 4 of this volume should be enough to get started. For example, let's say that your unknown is a common tree in a hedgerow. Let's say its name is *guásima* (pronounced wa' see ma). In this book you will find that guásima (*Guazuma ulmifolia*) is a tree of secondary communities that is used for making handcrafted furniture and must be harvested at a certain phase of the moon, according to some local carpenters.

One can also work the magic the other way around. Ask someone who lives in Aduana or some vaquero to direct you to whatever tree strikes your fancy, for example, palo santo. *"¿Donde hay un palo santo? Quiero verlo."* ("Where is a palo santo? I wish to see one.") Yes, a little Spanish helps. After all, TDF in North America is essentially confined to Spanish-speaking countries. The touchstone is that common name, which is common to the people native to this region. And don't worry. Vaqueros think that it is normal to be interested in plants of the monte. That is an attitude one does not often find by the side of the road north of the border.

Around Alamos and elsewhere, the forest improves as the roads get worse. In the foothills of the sierra, the country roads end at small ranches, each the head of a nexus of foot trails. Some of these lead into the heart of the country, the deeply dissected headwaters of the Río Cuchujaqui, or the barranca-rimmed tributaries of the Río Mayo. Narrow ravines, box canyons, and boulder-choked gorges defy even a superficial biological survey. Such rich habitats offer scant hope of a definitive study of litter production, nutrient cycling, seed dispersal, and other fundamental information about TDF that is now available mainly from one research station, the Biosphere Reserve in Chamela, Jalisco, more than 1,000 km to the south (see chapter 2), where the land is not so rough.

Dry-season plunge pools fed by modest trickles of water in narrow canyon bottoms between towering cliffs are too small to be resolved on aerial photographs. Moreover, the 1:50,000-scale CETENAL topographic maps typically indicate that a particular drainage in a canyon bottom is seasonal or temporary when in fact it is permanent, supporting a ribbon of unusual riparian trees in the shade of towering cliffs. In pitching canyons at elevations of 800 to 1,000 m in places such as Tepopa and Arroyo Verde, just below or interfingering with the land of oaks and pines, one finds the least known biota of Sonora and Chihuahua, including at least one new species of leguminous tree. The known species of trees include *Aphananthe* (*Mirandaceltis*; guasimilla), *Drypetes* (palo verde), *Magnolia pacifica* (ciprés), *Persea podadenia* (laurel, laurel cimarrón), and *Thouinia* (guasimilla roja), among others (Gentry 1942; Martin et al. 1998).

For a comprehensive overview of what eastern Sonora and western Chihuahua have to offer and of what is happening in this region, we recommend overflights in a light plane. In the mid-1990s an hour's round-trip flight from the Alamos airport east between Choquincahui and the Cerro Tarahumara and back over Rancho San Pedro revealed an almost continuous canopy of mature tropical forest, with clearings accounting for less than 5% of the land surface. In the summer rainy season the forest is fully foliated, a solid blanket of green, Sonora's "green mansions." We have our doubts about how long the green mansions will last. The drive into Alamos from Navojoa presents a sobering view of what has happened since our first trips in the 1960s. Only steepness limits the new clearings creeping up the slopes of the Sierra de Alamos.

At the May 1997 international conference on *desertification* (the loss of xeric grasslands to drought or overgrazing) held in Tucson, Arizona, Alejandro Castellanos of DICTUS (Departamento de Investigaciones Científicas y Técnicas de la Universidad de Sonora in Hermosillo) startled some of the delegates with a problem they had not heard of before. He lamented the fate of dry tropical forest and desert shrubs and cacti in Sonora, and he called it *grassification*. Ancient ironwood trees *(Olneya tesota)*, mature mesquite *(Prosopis glandulosa* and *P. velutina)*, palo santo, great candelabras of organpipe cactus, and dozens of other species of trees and shrubs have been bulldozed into windrows or *chorizos* and burned, or the uprooted mesquite and ironwood have been converted into charcoal and exported. Ecologists have long railed at such a fate. "To destroy any more ecological order, deliberately, in the interest of propagating cattle, would seem—no stronger word is helpful—simply daft" (Deevey 1967). Obviously the propagators of cattle see matters differently. Initially beef yields increase up to four times with buffelgrass.

Ranchers do heed market forces, and cattle ranching is no longer as attractive an investment opportunity as it once seemed. NAFTA (the North American Free Trade Agreement) has placed Sonoran ranchers at a comparative disadvantage. The Texan alternative, a game ranch stocked with exotic African or Asian bovids and cervids along with North American elk or bison, is being tried in Mexico, including on Rancho Tres Marías along the road to Alamos and on more than 20 other ranches in Sonora.

Livestock are latecomers in the New World, only recently being involved in natural selection for plant defenses (if one assumes that thorns and oils protect plants from being overeaten by megaherbivores). Native large animals that might be responsible for the evolution of thorns and oils include gomphotheres (the proboscidean genus *Cuvieronius*), glyptodonts, ground sloths, and native species of horses. According to Janzen and Martin (1982), these animals, which became extinct within the last 12,000 years, were also the original dispersers of hard seeds buried within carbohydrate-rich pods or fruits highly attractive to large mammals and some birds. Examples include chírahui, mesquite, and guamúchil *(Pithecellobium dulce)*. In the absence of extinct megafauna, people of the monte may serve as agents of dispersal for the seeds of the palatable fruits (Nabhan 1989).

How does Sonoran TDF compare with that found elsewhere in the Americas? Most if not all ecologists lack the field experience to attack the question on other than the most general terms. Even the obvious comparison between TDF in northeastern Mexico at 23°N latitude and that in the northwestern part at 27°N is neglected because hardly any ecologists, either Mexican or foreign, have spent much time in both areas. There are some striking similarities and intriguing differences between the two regions.

Because southern Sonora near Alamos receives less rainfall (500–900 mm) than southern Tamaulipas (900–1,200 mm), Sonoran trees are smaller. Both forests support a dense green canopy during the season of summer rains, but the season of full foliage is much shorter in Sonora. Both are largely leafless during their flowering seasons, winter and spring. Except during the onslaught of El Norte, the winters in southern Tamaulipas are hot and dry, unrelieved by the cool frontal rains or *equipatas* of southern Sonora. In the wettest of Sonoran winters, the gentle equipatas intensify into widespread torrential storms, which unleash raging floods that burst out of the sierra, overtop reservoirs, damage bridges, and inundate the fragile homes of thousands of people in the deltas of the Mayo and Yaqui Rivers. Matching them in severity are the fall hurricanes that blow down forest trees and devastate settlements on both Atlantic and Pacific slopes. In September 1995, Alamos received more than 50 cm (20 in.) of rain from Hurricane Ismael.

In southern Tamaulipas, TDF supports tropical tree genera that are scarce or unknown in Sonora and Chihuahua, such as *Amyris, Beaucarnea, Brosimum, Phoebe*, and *Ungnadia* (Puig 1976). Although Sonora features plenty of *espinal* (mesquite, chírahui, and other spiny trees), its TDF is more open, lacking the spiny thickets of the ground bromeliad *Bromelia pinguin* (huapilla) and other species that make up the impenetrable tangles often found in TDF in southern Tamaulipas. Although columnar cacti such as *Acanthocereus* and *Stenocereus* are not important forest elements in Tamaulipan TDF as they are in Sonoran TDF, the Tamaulipan tropical lowlands support the lianoid cactus *Pentacereus* and the epiphytic cactus *Rhipsalis*. Sonoran TDF features more species of the pleasantly odorous genus *Bursera*, although *Bursera simaruba*, the wide-ranging gumbo limbo of Florida, the West Indies, and most coastal lowlands within the tropical parts of Mexico, is relatively rare in Sonora.

Riparian trees lining drainages in both southern Sonora and Tamaulipas include stately sabinos *(Taxodium)* that can exceed 30 m in height and 1.5 m in diameter and typically harbor poison ivy *(Rhus radicans)*. Roadside hedgerows in both provinces support guásima. Genera of tropical trees in Sonoran TDF that are scarce or unknown in Tamaulipas include *Albizia, Cochlospermum, Conzattia*, and *Haematoxylon. Conzattia*, known locally as *joso* and the tallest tree in the forest, is a West Coast endemic. Although Sonoran and Sinaloan TDF and thorn forest formations are rich in endemics, similar habitats in southern Baja California are even richer (Turner et al. 1995). Tamaulipan TDF lacks the shrubby wind-pollinated ambrosias (chicurilla) so important in disturbed TDF in Sonora. On nutrient-poor soils in both regions, oaks *(Quercus)* may insert themselves into the tropical lowlands. For example, *Q. olioides* (encino) occupies sterile shales in the lowland forests of eastern Mexico, whereas various species of oaks, especially the cusi, *Q. albocincta*, appear on barrens of acid soils formed by altered outcrops within TDF of western Mexico (Goldberg 1982).

An outstanding example of a red altered and mineralized outcrop may be found near Alamos around the former Mayo town of Piedras Verdes. Here *Q. albocincta* grows at the unusually low elevation of 250 m, 500 m below its usual lower limit in the sierran foothills. The Piedras Verdes outcrop caps a rich ore body that has recently been extensively core-drilled and apparently is destined for major exploitation. The unique plant associations limited to the red soils may be lost.

Both southern Tamaulipas and southern Sonora host a variety of tropical birds: chachalacas, motmots, squirrel cuckoos, trogons, lineated woodpeckers, woodcreepers, tropical robins, and many species of parrots. Tamaulipas is home to brown and green jays, whereas southern Sonora harbors the black-throated magpie jay and purplish-backed jay. Tinamou *(Crypturellus)*, shy grouse-sized game birds, haunt tropical forests of southern Tamaulipas. Although bird watchers in southern Sonora will find many similarities to the avifauna of the southern United States, the Sonoran forests and riparian areas contain more tropical species, such as chachalacas, cotingas, motmots, parrots, tiger herons, and trogons (see chapter 7).

TDF is home to interesting reptiles as well. Southern Tamaulipas features the bizarre spindly legged, tropical lizard *Laemanctus* and the venomous snake known as *cuatro narices (Bothrops atrox)*. In Sonora the venomous, stumpy-legged, beaded lizard *Heloderma horridum* lives in tropical habitats, as well as the elusive, venomous crotalid snake *Agkistrodon bilineatus*. Known as the *pichicuate*, it is rarely found in Sonora (see chapter 6).

Among the more remarkable mammals that both regions support are mouse opossums *(Marmosa)*, vampire bats *(Desmodus)*, jaguar *(Panthera onca)*, and mountain lions *(Puma concolor)*. Legends of the *onza* (neither a jaguar nor a mountain lion) are rife in the sierran foothills of Sonora. Southern Tamaulipas is the northern limit of tayra *(Eira)*, which is a 5-kg tropical mustelid, and of brocket deer *(Mazama)*.

What about the people who live in these places? Although many have lost their roots in the ferment of contact and economic transformation, TDF in southern Tamaulipas and eastern San Luis Potosí remains the traditional domain of the Huastecans, a horticultural people who possess a rich knowledge of their native tropical biota. Those on intimate terms with TDF in Sonora and Chihuahua include at least four distinct groups, the Guarijío (Makurawe), Yoreme (Mayo), Tarahumara, and Yoeme (Yaqui), plus assimilated peoples from earlier times such as the Chínipas and Basiroas. Although in jeopardy, the ethnobotanical and other traditional knowledge of these groups endures.

We will illustrate what we mean with a few examples. Research with our colleagues, Rigoberto López Estudillo of the University of Sonora and Tom Van Devender of the Arizona-Sonora Desert Museum, indicates a variety of economically useful plant products that can be harvested from TDF, foothills, coastal thorn forest, and thornscrub.

Etcho is a columnar cactus that is dominant in the lower parts of TDF near Alamos. *Pachycereus pecten-aboriginum* yields a delectable fruit. The seeds, produced in great abundance, are made into *atole* (a mush similar to cream of wheat) and excellent tortillas. The juice of the trunk is used for healing cuts and insect stings. The ribs are valuable in light construction. David Yetman recently asked one of his Mayo consultants if he had ever eaten a tortilla made of etcho seeds. The man answered, with a hint of indignation, "Have I eaten it? I grew up on it!"

For six weeks in early summer the organpipe cactus, or pitahaya, yields sizable amounts of a sweet and much sought fruit. Dried pitahaya, called *pitahaya seca*, has excellent nutritional components as well as alleged aphrodisiac powers. The ribs of the dead cactus are used in light construction, including fences, beams, ceiling lattices, and several household implements.

No fewer than 20 woods—including amapa *(Tabebuia impetiginosa)*, palo blanco, palo colorado *(Caesalpinia platyloba)*, mauto, guayacán *(Guaiacum coulteri)*, vara blanca *(Croton fantzianus)*, and batayaqui *(Montanoa rosei)*— are used for a wide variety of construction purposes. Native peoples have long made their climatically appropriate homes entirely from local materials. The amapa is unmatched for strength, resistance to rot, and insect damage. Palo colorado is so strong and durable that it is said to produce fence posts that will endure for 100 years.

At least 10 species of trees yield renewable firewood. Probably half the population of southern Sonora relies on firewood or charcoal for cooking.

Many legume species are important for fodder and the shade they provide livestock. Livestock eat the pods and, we are told, humans even eat the supposedly highly toxic beans of the chilicote. During the prolonged dry cycle of 1995–1996, péchita (mesquite pods) were the main livestock fodder after grass and forbs were consumed.

Many native plants are used for medicinal purposes. Several have been used for such a long time that it is hard to doubt their efficacy. The lac produced by insects called *goma de Sonora* on *Coursetia glandulosa*, locally known as *causamo* and found in tropical forests near Alamos, is praised throughout Mexico as a broad-spectrum remedy. One of our Mayo consultants stores bags of bark from five different trees and shrubs as medicine for common ailments. We watched him one day hack away bark from the aerial roots of the rock fig, tescalama *(Ficus petiolaris)*, which, he reported, is a good cure for hernia. The bark of two handsome trees, palo mulato *(Bursera grandifolia)* and copalquín *(Hintonia latiflora)*, is routinely used and marketed throughout southern Sonora. Copalquín "shows significant antimalarial and antidiabetic activity" (Turner et al. 1995).

Many trees and shrubs provide wood suitable for homemade artifacts.

Güiloche *(Diphysa occidentalis)* is widely used for canes and tools and was presented as a splint to a member of David Yetman's party who suffered a mildly sprained ankle on a trek into the Sierra de Alamos. From a variety of local woods, Mayo and Guarijío alike produce masks, drums, spears, and musical instruments for both the tourist trade and home use. Fast-growing palo joso *(Albizia sinaloensis)* can be carved into many kinds of utensils, bowls, and trays that will last much longer than more fashionable plastic or ceramic dishes.

We have little time to find out what uses remain. The older people, the keepers of the lore, will not last much longer. Youths do not find that the ancient pathways to the knowledge of their elders are taught in school or venerated on TV. Although we cannot verify all that we are told, our one regret is that we do not know the environment and ethnobiology of the region well enough to ask the right questions so that we might learn more from those who live there.

The "buffelgrass bomb" (Bowden et al. 1993) presents a cultural as well as an ecological dilemma. We approach the problem from a simple case history, the *niche* in ecological jargon, of a man named Marcelino Valenzuela Cota, a Mayo (fig. 1.1).

Born in 1949, Marcelino has lived in the area of Alamos, Sonora, all his life. He was orphaned and raised by distant relatives, but spent enough time in the monte to become perfectly at home in these forests that are removed from most conveniences of modern technology. For the last few decades, his livelihood has come from two sources: as a master chair maker in the Mayo tradition (Krizman 1968) and as a woodcutter of vara blanca.

With simple tools he cuts limbs from the guásima tree. He does this only during the full moon because he finds the wood too soft and brittle at other times. He hauls the limbs by burro to his tiny home near a ravine in a barrio of Alamos or to his family's *ranchito* on the *ejido* (community-owned lands) at El Rincón a few kilometers to the north. With hand tools he painstakingly rips the wood into planks of the appropriate thickness and saws and planes the lumber into the chair pieces. With an expert's eye he fashions mortises and tenons and fits the chair pieces together, relying on tight-fitting joints more than nails to secure the chair's frame. He uses nails primarily to fasten the slats of the seat. The finished product is a marvel of design and function and will last a good 20 years of hard use (fig. 1.2). His cousin has such a chair that is more than 40 years old. Marcelino's chairs are snatched up by buyers as fast as he can make them.

FIGURE 1.1. Marcelino Valenzuela Cota of El Rincón, Sonora. Photograph by Thomas R. Van Devender.

Marcelino, along with many others, also harvests vara blanca, one of the major forest products of TDF in southern Sonora. Vara blanca is the ideal size and taper for tomato stakes and is delivered in many truckloads to irrigated farmland in Baja California and elsewhere.

Marcelino's way of life and that of hundreds of others like him demand sizable areas of intact TDF near inhabited areas. These forest areas constitute a refugium of materials and resources—firewood, lumber, medicines, dyes, fence posts, shade, forage for livestock, and so on—without which Marcelino's semisubsistence way of life could not be maintained. Disappearance of the forest threatens both individual livelihoods and the very survival of rural populations. As the forests dwindle, so do these necessities, these hidden props that long ago invited and allowed ongoing human presence.

Neither *vacas* (cattle) nor limited conversion to buffelgrass pastures per se are a threat to the forest. Vacas have ranged the monte since early colonial times. Clearings in the forest have come and gone for centuries. Even today in the rainy season, the uniform light green of a successional species, the chírahui, reveals the location of ancient milpas (locally known as *mahuechis*). The threat also does not lie with the small rancher who carries out the *desmonte* (cutting and clearing) of a few dozen hectares. The aggressive buffelgrass meets its match on steep, rugged mountain slopes covered by TDF. Even on level sites it cannot invade the intact forest without human assistance and disturbance. Chírahui, vinorama *(Acacia farnesiana)*, pintapán *(Abutilon abutiloides)*, and chicurilla *(Ambrosia cordifolia)* are the advance guard of low trees and shrubs in forest recovery. With a supply of regional seed sources or stump sprouts, the trees will return. The new threat to the traditional compromise between humans and the forest comes from ominous socioeconomic trends occurring in the region: the rise of absentee cattle lords and the growth of *narcotraficantes*.

Small-time resident ranch owners with small numbers of livestock are becoming as rare in northwestern Mexico as in the United States. It appears to us that northwestern Mexico's political and social elite, no less than their U.S. counterparts, seek the stamp of authenticity brought by ranch ownership. As this small but powerful group has expanded, so has the desire of its adherents to acquire ranch lands. Throughout the Sonoran foothills one finds large ranches of recent acquisition by wealthy members of upper classes from Hermosillo, Ciudad Obregón, and Navojoa. Management of the ranch is turned over to a local *ejidatario* or vaquero. The new owner visits once or twice a month to supervise, to ride a horse, and to count the number of livestock. He does not live on the ranch and has no feel for the traditional ranch life. The important thing is that his social associates will know that he owns a ranch. Beyond that he only wishes to show a profit. The quickest return on a ranch investment is to clear the land, sow buffelgrass, and see the beef production soar. The welfare of the land may be secondary to the return. There are numerous 1,000-ha clearings of buffelgrass in southeastern Sonora that are the result of such absentee wish fulfillment and accounting dictatorship.

We have reason to believe there is a darker side to the spread of some of the new clearings at the ends of new roads within expensive new barbed-wire fences. In this way *mafiosos* launder money from the drug trade. Government figures are obviously unavailable to quantify the trend. Interviews with rural folk readily establish the prevalence of the practice. Old-time ranchers and small and medium landholders find their parcels coveted by mafiosos. In some cases shocked ranchers will find newly irrigated plots of marijuana

FIGURE 1.2. Chair made of guásima *(Guazuma ulmifolia)* by Marcelino Valenzuela Cota of El Rincón, Sonora (September 1994). *A, above*, chair; *B, opposite*, detail of joint. Photographs by Thomas R. Van Devender.

on their land, each watched over by an armed thug. Protest is folly; the *narcos* have the upper hand. Acquiescence also is dangerous; the army may find the new patch on their next annual drug sweep. Then an offer is made to the rancher, which, under the circumstances, is hard to refuse. In this way, ranch by ranch, the mafiosos penetrate the sierra. Once they have assembled a large spread, they clear all of it, flats and hills alike, plant buffelgrass, and introduce as many vacas as possible. This new enterprise is a convenient means of accounting for otherwise inexplicable incomes and expenditures.

The forest thus faces two potent adversaries: the concentration of Mexico's wealth in an urban elite and the proliferation of rich mafiosos. Through these dual forces, which sometimes are one and the same, tens of thousands

B

of hectares have been cleared in the valley of the Río Mayo above Mocúzari Dam, in the foothills of the Río Yaqui drainage, and in the Río Cuchujaqui and Río Fuerte basins. Our meanderings through the region have brought us numerous stories of traditional ranches lost to one or the other or both, of subtle pressures, of outright extortion, of political favors, and of mortal threats.

There may be a way to turn the tide. In the mid-1990s the World Bank considered investing $60 million in a scheme to improve forestry in the Sierra Madre Occidental of Chihuahua and Durango. The project was abandoned, some suspect, when the bank became aware of the environmental consequences of its proposal. We have a new proposition for the bank or anyone else with deep pockets and a green vision. For considerably less than $60 million, it should be possible to secure a sizable tract of tropical deciduous forest, a valuable array of traditional crop plants along with their genetic variability, and the people with deep roots in this region, who will lose their ecological capital if the forest is grassified, privatized, or otherwise sacrificed and

taken from them. Thanks to the roughness of the terrain on the skirts of the
sierra, many patches of tropical deciduous forest should be safe, even from
the buffelgrass bomb. Make no mistake: This is a land to save and to savor.

REFERENCES

Allen, W. H. 1988. Biocultural restoration of a tropical forest. *BioScience* 38:156–161.

Bowden, C., J. W. Dykinga, and P. S. Martin. 1993 *The secret forest*. Albuquerque: University of New Mexico Press.

Deevey, E. S., Jr. 1967. Introduction. In *Prehistoric extinctions*, eds. P. S. Martin and H. E. Wright, 63–73. New Haven: Yale University Press.

Douglas, M. W., R. A. Maddox, and K. Howard. 1993. The Mexican monsoon. *Journal of Climate* 6:1665–1677.

Gentry, H. S. 1942. *Río Mayo plants: A study of the flora and vegetation of the valley of the Río Mayo, Sonora*. Publication No. 527. Washington, D.C.: Carnegie Institution of Washington.

Goldberg, D. E. 1982. The distribution of evergreen and deciduous trees relative to soil type: An example from the Sierra Madre, Mexico, and a general model. *Ecology* 63:942–951.

Janzen, D. H., and P. S. Martin. 1982. Neotropical anachronisms: The fruits the gomphotheres ate. *Science* 215:19–27.

Krizman, R. D. 1968. Manuel Vaquasewa, a chair maker from Alamos, Sonora, Mexico. *Keystone Folklore Quarterly* Summer:103–110.

Martin, P. S., D. A. Yetman, M. E. Fishbein, P. D. Jenkins, T. R. Van Devender, and R. Wilson, eds. 1998. *Gentry's Río Mayo plants: The tropical deciduous forest and environs of northwest Mexico*. Tucson: University of Arizona Press.

Nabhan, G. P. 1989. *Enduring seeds: Native American agriculture and wild plant preservation*. San Francisco: North Point Press.

Puig, H. 1976. *Vegetation de la Huasteca, Mexique*. Etudes Mesoamericaines, vol. 5. México, D.F.: Mission Archeologique et Ethnologique Francaise au Mexique.

Turner, R. M., J. E. Bowers, and T. L. Burgess. 1995. *Sonoran Desert plants: An ecological atlas*. Tucson: University of Arizona Press.

2

STRUCTURE AND FUNCTIONING
OF TROPICAL DECIDUOUS FOREST
IN WESTERN MEXICO

Angelina Martínez-Yrízar, Alberto Búrquez,
and Manuel Maass

Tropical deciduous forest (TDF) is distributed over large areas in the dry, markedly seasonal, tropical regions of the world (Murphy and Lugo 1986, 1995). On the North American continent, TDF extends from northern Panama to northwestern Mexico. TDF (*monte mojino* [Ochoterena 1923], short tree forest [Gentry 1942], *selva baja caducifolia* [Miranda and Hernández-X. 1963], or *bosque tropical caducifolio* [Rzedowski 1978]) thrives throughout Mexico under seasonal A-type climates (García 1973) and from sea level to about 1,600 m in elevation. TDF is distributed between the Pacific Coast and the foothills of the Sierra Madre Occidental and Sierra Madre del Sur, from east-central Sonora and western Chihuahua to Chiapas. It also grows in mountain ranges in Baja California Sur, the Río Balsas Basin, southern Tamaulipas, southeastern San Luis Potosí, and the Yucatán Peninsula (Rzedowski 1978; fig. 2.1).

Despite its geographic and cultural importance, TDF is among the least known tropical ecosystems. It is also considered the most endangered tropical forest ecosystem because of the high rate of deforestation and the scarcity of protected areas for conservation (Flores-Villela and Geréz 1988; Gentry 1995; Janzen 1988; Maass 1995). According to Jaramillo-Villalobos (1994), TDF constitutes the most extensive area of tropical forest vegetation in Mexico (ca. 15.6 million ha). In the last 50 years, it has been rapidly transformed into induced grasslands and agricultural fields, causing a decline in biological diversity and a disruption in ecosystem processes (De Ita-Martínez 1983; Maass 1995; Yetman and Búrquez 1994). Reduction in vegetation cover due to forest conversion seriously changes water and nutrient cycles, increases soil erosion, and modifies the properties of the soil (Jaramillo-Luque 1992; Maass

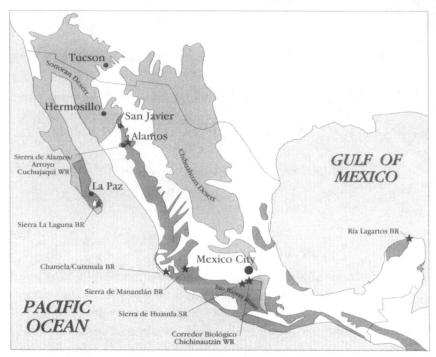

FIGURE 2.1. Distribution of tropical deciduous forest in North America. BR = Biosphere Reserve; WR = Wildlife Refuge. From Búrquez et al. (1999), INEGI (1998), and Rzedowski (1978).

1995). Once the topsoil is removed by erosion, opportunities for forest restoration are highly reduced (Lal 1990; Maass et al. 1988).

Tropical forests have been recognized as rich reservoirs of resources such as food, building materials, fuel, tools, and medicines (Gentry 1942; Lumholtz 1902; Toledo 1988). However, only a few ethnobotanical studies in Mexico have focused on dry forest areas (Bye 1995; Gentry 1942; Yetman et al., this volume). The traditional uses of TDF plants have been recorded in the Río Mayo region in Sonora by David Yetman and his collaborators. They have documented the multiple uses of the ecosystem by Mayo natives (Yetman et al., this volume). As the first step in medical treatment, native Mayo housewives use about 90 species of native tropical forest and thornscrub plants (Bañuelos and Búrquez 1996). One can only imagine the vast number of plants used by knowledgeable native *curanderos* (medicine men). Examples of the ethnoecology of specific plants are few, but some plants are better documented, such as the palms in Sonora, especially the traditional ecological knowledge and resource management of *Sabal uresana* (palma real; Joyal 1996).

In this chapter we will describe the structural and functional attributes of TDF along the Pacific coast of Mexico. Emphasis will be placed on the forest near Chamela, Jalisco, because it is pristine and extensive and probably represents the most studied TDF in western Mexico. Wherever possible, we compare the Chamela TDF with the forests in Alamos and San Javier, Sonora, where we have started comparative data collection.

Studies in Mexico

Studies regarding the structure, floristic composition, and ecosystem functioning of TDF in western Mexico include the Río Mayo Basin (Gentry 1942), the region of Nueva Galicia (central Sinaloa to Colima; Rzedowski and Mc-Vaugh 1966), Sierra La Laguna (Arriaga and León de la Luz 1989), the Río Balsas Basin (Miranda 1947), and Chiapas (Miranda 1952). The Oaxacan (Rzedowski 1978) and Tamaulipan TDF are the least surveyed.

Three regions have prominence among TDF studies: (1) the Río Balsas Basin, (2) southern Sonora, and (3) the Chamela region. The Río Balsas Basin, perhaps the richest reservoir of TDF vegetation in Mexico, has been studied intensively from a floristic standpoint (Dávila et al. 1993; Gual 1995; Kohlman and Sánchez-Colón 1984; Miranda 1947; Peralta 1995; Rzedowski 1978; Vargas and Pérez 1996). In southern Sonora, extensive botanical surveys have been conducted in the Río Mayo region (Gentry 1942; Martin et al. 1998; Van Devender et al., this volume; Yetman et al. 1995). Aspects such as the influence of physical environment and the distribution and phenology of animal and plant species have also been reported in several localities in Sonora (Búrquez 1997; Krizman 1972; Wiseman 1980). Goldberg (1982, 1985) has shown the substantial effect of soil mineralization and acidity on the distribution of low-elevation oak woodland as well as pine-oak woodland and TDF.

Most of the work at the ecosystem level has been carried out at the Estación de Biología Chamela, a nature reserve of the Universidad Nacional Autónoma de México in Jalisco (Sarukhán and Maass 1990). Current research that follows a more global approach contrasts the carbon pools with the dynamics of intact and disturbed, tropical, humid and deciduous forests in Mexico (Kauffman et al. 1994). We have recently started research in the gradsect along Mexico's Highway 16 (Búrquez et al. 1992a), which includes sizable areas of TDF. Our main aims focus on the composition and structure of the vegetation, land-use change, carbon balance, and ecosystem dynamics.

Physical Environment

The warm isothermal climate of the Chamela region is influenced by tropical cyclones that produce a highly variable, annual rainfall regime (Bullock 1988; García-Oliva et al. 1991). Mean annual precipitation is 696 mm (1982–1995), but the difference between the wettest and the driest year is more than 800 mm (1,186 mm in 1992 and 374 mm in 1985). Rainfall distribution is strongly seasonal: 80% of the annual precipitation falls from July to October. September is the wettest month, with about 160 mm. This weather pattern is prevalent along the Pacific coast of Mexico up to east-central Sonora. The macroclimate in the Sierra de Alamos in southern Sonora is as warm and strongly seasonal as in Chamela. Rainfall from June to October accounts for 85% of the total annual precipitation (639 mm). The mean annual temperature is slightly cooler than in Chamela (23.5°c vs. 24.9°c; Krizman 1972).

The west-to-east gradsect in central Sonora exhibits dramatic changes in elevation and climate on the passage from the coast to the Sierra Madre Occidental (Highway 16). In a short distance, the climate changes from the steppe BS and BW types almost at sea level in Hermosillo, to the tropical Aw type in San Javier at the 550–1,100-m elevation, to the humid temperate Cw type high in the sierra in Yécora at the 1,900-m elevation. (For climate types, see García 1973.) The onset and amount of annual rainfall and the mean annual temperature thus differ markedly along this gradient, causing striking changes in vegetation. This is the richest gradsect in the Pacific Coast of North America. TDF is usually found below the 1,200-m elevation, intergrading sharply with oak woodland above and imperceptibly with thorn forest below at about 500 m (Búrquez et al. 1992a, 1999; Gentry 1942). TDF at 29°N latitude is pressed against the foothills of the Sierra Madre Occidental, about 250 km inland, surrounded by desert and thornscrub. Southward the desert-and-thornscrub band narrows as the sierra approaches the sea until, near the tropic of Cancer, TDF develops near the shore. Gradsects near Mazatlán, Sinaloa, lack the rich lowland bands of Sonoran Desert and thornscrub, whereas gradsects north of Hermosillo, Sonora, lack the diversity and structure of TDF.

Although dry forests are primarily climatically determined (Murphy and Lugo 1995), soil properties and topography can either constrain or enhance the local distribution of TDF within a larger homogeneous climatic matrix (Búrquez et al. 1999; Goldberg 1982, 1985). In Chamela, soils at the base of the hills, in valleys, and along the river channels show a higher organic-matter and moisture content than their counterparts on the slopes (Galicia 1992; García-Oliva and Maass 1998). On these more mesic sites, TDF is displaced by a taller, denser, and more productive semideciduous forest (Martínez-Yrízar and Sarukhán 1990). In contrast, at the Sierra Libre and Sierra El Aguaje, Sonora, 10° northward in latitude from Chamela, because of the scant desert precipitation TDF species are found at the bottom of the deep canyons and ravines of the well-delimited Sonoran Desert mountain ranges (Búrquez et al. 1999; Felger 1999; Yetman and Búrquez 1996). On these sites, the combined effect of microclimate (with higher relative humidity and lower potential evapotranspiration) and the extended and increased water availability through seeps in the massive ignimbrites create an environment where most of the structural and functional attributes of TDF persist.

Forest Structure

The structure of tropical dry forests varies enormously from site to site. Murphy and Lugo (1986) listed the typical structural and functional properties of these forests. Among the major attributes, they included (1) canopy height from 10 to 40 m, (2) basal area of trees from 17 to 40 $m^2 ha^{-1}$, (3) total biomass from 98 to 320 $Mg ha^{-1}$, (4) root biomass about 50% of total biomass, and (5) tree species richness from 30 to 90 species in relatively small areas (less than 3 ha). TDF, being the most xeric subset of the tropical dry forests, exhibits structural parameters in the lower limits of the aforementioned ranges.

The first modern botanists to formally describe the structural and functional attributes of TDF were Howard Scott Gentry (1942), who studied the Río Mayo Basin in southeastern Sonora, Mexico, and Faustino Miranda (see Rzedowski 1978), who studied different localities in central Mexico. Apart from the striking transformation of the forest from verdant full-leaf to gray leafless trunks, Gentry recognized as a diagnostic feature the relatively short height of the trees and baptized this ecosystem the *short-tree forest*. According to Yetman et al. (1995), two important characteristics of the Sonoran TDF are the dominance of mauto (*Lysiloma divaricatum*), which typically grows in nearly pure stands, and the abundance of *Bursera* species (eight in the Sonoran TDF; Johnson 1992). These traits are not restricted to Sonora, however; forests in Sinaloa, Nayarit, and Colima have the same dominants (Gentry 1946; Rzedowski and McVaugh 1966).

In the Río Balsas Basin, TDF contains a multitude of sympatric species of *Bursera* as codominants (in a locality in Guerrero, as many as 27 different species; Kohlman and Sánchez-Colón 1984), along with *Lysiloma divaricatum, Tabebuia impetiginosa* (amapa), and many others, including several columnar and arborescent cacti (Rzedowski 1978; Rzedowski and McVaugh 1966). Additional structural attributes of TDF in the Río Mayo region are its variable but unbroken 9- to 15-m canopy; the mixed perennial population in typical stands numbers 1,000–2,000 individuals per hectare (ind. ha^{-1}; Gentry 1942). Other studies in the same region offer similar values. Quadrat analysis used by Krizman (1972) to describe the plant community in a 524-m-elevation site on the south side of the Sierra de Alamos showed that tree density was about 1,050 ind. ha^{-1}. At this site, *Brongniarta alamosana* (palo piojo), *Lonchocarpus hermanii* (nesco), and *Pachycereus pecten-aboriginum* (etcho) were the most common tree species.

More recent vegetation sampling in the Sierra de Alamos, using the point-centered quarter method, shows that dominance, density, and basal area of trees vary greatly from site to site in relation to slope orientation and steepness. In a somewhat disturbed site on a gentle slope, *Croton alamosanus* (vara blanca) was the most common species; the total estimated density was 1,702 ind. ha^{-1}. In a 450-m-elevation site on a gentle east-facing slope that had been slightly grazed by cattle, the mean density was 1,900 ind. ha^{-1}; *Alvaradoa amorphoides* (palo torsal), *Caesalpinia caladenia* (margarita), *Jatropha malacophylla* (sangrengado), and *Pachycereus pecten-aboriginum* were the most abundant species. At a nearby site (El Tempisque), *Brongniarta alamosana, Erythroxylon mexicanum* (mamoa), *Montanoa rosei* (batayaqui), and *Wimmeria mexicana* (algodoncillo) were dominant. In northwest-facing slopes, density reached about 1,772 ind. ha^{-1}. The basal area of trees differed between sites by more than one order of magnitude, with the lowest values on northwest-facing slopes and the highest on east-facing slopes (15 and 40 m^2 ha^{-1}, respectively; table 2.1).

Near its northern distributional limit, TDF is found in San Javier, Sonora, at elevations of 550–1,100 m (Búrquez et al. 1992a, 1999; Wiseman 1980). Here the forest also differs in floristic composition and structure in relation to elevation, soil composition, and slope aspect and steepness. For example, stands of *Tabebuia impetiginosa* are more common on hydrothermally altered soils (Paul Martin, pers. comm. 1992). The vegetation on south-

Table 2.1. Structural Attributes of TDF in Western Mexico

Site	Density (ind. ha^{-1})	Basal Area (m^2 ha^{-1})	Biomass (Mg ha^{-1})	N	Sample Units	Slope A	Slope S(°)	Elev. (m)
SAN JAVIER (28°36'N, 109°44'W)								
San Juan 1[a]	1,550 (607)	20.1 (9.5)	58.4 (27.7)	16	50 × 2	NE	30	1,000
San Juan 2	2,120	28.1	81.5	22	PCQ	NE	35	900
Corrales[a]	1,425 (500)	26.7 (13.6)	77.7 (39.5)	16	50 × 2	NE	34	550
Mirador[a]	1,813 (363)	17.5 (9.4)	50.2 (27.2)	16	50 × 2	SE	34	800
ALAMOS (27°01'N, 108°56'W)								
Vara Blanca	1,702	20.7	60.3	16	PCQ	Level	> 10	nd
Tempisque	1,772	15.4	44.8	29	PCQ	NW	10–30	500
La Luna	1,900	40.4	117.5	46	PCQ	E	> 10	450
CHAMELA (19°30'N, 105°03'W)								
Watershed I[b]								
Upper	2,790	12.7	37.0	1	80 × 30	All	8–16	150
Middle	3,221	17.3	50.4	1	80 × 30	All	8–16	130
Lower	2,104	19.8	57.7	1	80 × 30	All	16–30	70
Hill[c]	2,140	25.6	81.0	1	50 × 20	S	20–25	90

Notes: Density, basal area, and aboveground biomass include all stems greater than 3.0-cm diameter at breast height. Biomass values were predicted from the regression equation obtained by Martínez-Yrízar et al. (1992). N = number of samples. Sample units = quadrats (m) or PCQ (Point Centered Quarters). A = slope aspect; S (degrees) = steepness. Elev. = elevation; nd = not determined. Standard variation of the mean is given in parentheses where available.
[a] Boone Kauffmann, Victor Jaramillo, and Angelina Martínez-Yrízar, unpubl. data.
[b] Martínez-Yrízar et al. (1996).
[c] Martínez-Yrízar et al. (1992).

facing slopes includes these dominant species: *Bursera lancifolia* (torote copal), *B. laxiflora* (torote prieto), *B. penicillata* (torote acensio), *Hintonia latifolia* (copalquín), *Jatropha cordata* (papelío), and *Lysiloma divaricatum*. On low-elevation, northeast-facing slopes, *Haemathoxylon brasiletto* (palo brasil), *Croton flavescens* (vara prieta), and *Lysiloma divaricatum* predominate, whereas at higher elevations, *Brongniartia alamosana, Hintonia latifolia*, and *Senna atomaria* (palo zorrillo) are the most common species. Density varies from 1,400 to 2,100 individuals, and basal area ranges from 17 to 28 m^2 ha^{-1} (table 2.1).

In Chamela the forest is distributed in a series of low hills (mostly below 150 m) with slope gradients normally 21°–34° (Bullock 1988). Canopy height is typically 4–15 m. Total basal area ranges from 13 to 26 m^2 ha^{-1}, and average density from 2,104 to 3,221 ind. ha^{-1} (table 2.1; Lott et al. 1987; Alfredo Pérez-Jiménez, unpubl. data). Total aboveground living biomass ranges from 50 to 85 Mg ha^{-1}, with a root:shoot biomass ratio of 0.42 (Castellanos et al. 1991; Martínez-Yrízar et al. 1992). Topographic position appears to be related to differences in total aboveground phytomass, with higher phytomass in the lower slopes (Kauffman et al. 1994). A structural feature of the forest is that only one-third of the standing dead-wood biomass decomposes on the forest floor; the rest hangs, attached to living trees or dead standing trees (Manuel Maass, unpubl. data). Standing dead necromass ranges from 8 to 10 Mg ha^{-1} (Kauffman et al. 1994). Belowground biomass is 30 Mg ha^{-1}. Nearly two-thirds of all roots are confined within the 0- to 20-cm soil layer, and about one-third of all roots are less than 5 mm in diameter (Castellanos et al. 1991). Leaf Area Index—an important structural characteristic of the forest because the canopy is the site of significant ecosystem processes, such as transpiration, rainfall interception, dry deposition, and photosynthesis—fluctuates from 1 to 4.5 m^2 m^{-2} during the year (Maass et al. 1995).

In a study of floristic composition, Lott (1993) recognized 1,120 species of vascular plants in 124 families in the Chamela Bay region. Leguminosae is the predominant family (14%), followed by Euphorbiaceae (8%), Asteraceae (6%), and Graminae (5%). Similar values are found throughout the tropical forests along the Pacific slope of the Sierra Madre Occidental (Martin et al. 1998; Rzedowski and McVaugh 1966; Van Devender et al., this volume). In the Río Balsas Basin, Burseraceae is the fourth most important family of vascular plants. The presence of dense *cuajiotales* (*Bursera*-dominated forests) makes this region unique among the tropical deciduous forests of Mexico (Gual 1995; Miranda 1941; Peralta 1995; Vargas and Pérez 1996).

Forest Functioning

The total amount of solar energy converted by photosynthesis into chemical energy in plants is called Gross Primary Productivity (GPP). Some of the sugars that constitute this production are used by the plants for cellular respiration. The chemical energy that remains after the respiratory demands are met is called Net Primary Productivity (NPP). This energy may be stored as simple organic compounds (i.e., starch) in existing plant tissues or as specialized carbohydrates such as cellulose or lignin. NPP thus constitutes the total converted solar energy available for harvest by heterotrophs or consumers (i.e., herbivores) and for reduction by saprobes.

In any ecosystem NPP provides the energetic and material basis for the life of all organisms, not just the plants (Whittaker 1975). NPP is commonly measured in terms of dry organic matter synthesized per unit area per unit of time. This measure should be distinguished from the ecosystem biomass, which is the total amount of organic matter (above- and belowground) present at a given unit of time (Cox and Atkins 1979). NPP of terrestrial ecosystems depends on a variety of factors, including total solar radiation, temperature, rainfall, and the availability of essential nutrients. There is a general global pattern for biomass as well as for annual NPP: both decrease along moisture and temperature gradients. Departures from this global trend are related to local perturbation of the macroclimate by the presence of mountain ranges or the closeness to the sea.

Production measurements for TDF ecosystems are still scarce (Martínez-Yrízar 1995; Murphy and Lugo 1986). The Chamela forest is perhaps the only site that has been intensively studied; it has provided detailed information for more than 10 years. The aim of the Chamela project is to determine the long-term behavior of the primary forest (pools, energy flux, and nutrient cycling) and the responses of the ecosystem to different human exploitation practices. The productivity studies have included the quantification of functional processes, such as the production of litter (Martínez-Yrízar and Sarukhán 1990); surface litter and decay (Martínez-Yrízar and Sarukhán 1993; Patiño 1990); caloric content of the vegetation (Aznar 1993); levels of herbivory (Filip et al. 1995); and stem wood and fine root production (Kummerow et al. 1990; Alfredo Pérez-Jiménez, unpubl. data). With the integration of these parameters, the first direct estimate of total NPP for a Neotropical deciduous forest was determined (Martínez-Yrízar et al. 1996). Estimates at Chamela indicated that fine root production (4.2 Mg ha^{-1} y^{-1}), litterfall (3.6 Mg ha^{-1} y^{-1}), and aboveground wood increment (2.4 Mg ha^{-1} y^{-1}) constituted the most important components of total NPP (12.1 Mg ha^{-1} y^{-1}). These values are within the range of values for tropical dry forests reviewed by Martínez-Yrízar (1995) and Murphy and Lugo (1986).

Perhaps the most remarkable feature of TDF is the seasonality in litter production, which is coupled to the pronounced seasonal pattern of rainfall distribution. With few exceptions, the species remain leafless for several months during the dry season each year (Bullock and Solís-Magallanes 1990). In the Chamela forest, the rate of litterfall is maximum at the start of the dry period, and most of the annual litter is produced during the drought. Dry season litterfall (Nov.–Feb.) is 58–79% of the total annual value, which varied from 3.6 to 4.3 Mg ha^{-1} y^{-1} over a 5-year period (Martínez-Yrízar and Sarukhán 1990). Litter production in TDF also represents a more synchronized pulse of nutrient input to the soil system than that in many other tropical forests (Jaramillo-Luque and Sanford 1995). These nutrients are released during the process of litter decomposition, which takes place mainly during the short wet season. Decomposition experiments show that most (> 50%) of the leaf litter breaks down within the 3 months following the onset of the wet season, and little degradation occurs during the dry season (Martínez-Yrízar 1980).

From the coast of Jalisco, where the Chamela forest is located at 19°N latitude, large tracts of TDF extend northward to the Mexican states of Nayarit, Sinaloa, and Sonora. Here pockets of forest are found along the Río Sonora, Río Moctezuma, and Río Yaqui-Bavispe between 29° and 30°N latitude (Búrquez et al. 1999; Krizman 1972), reaching the northern distributional limit of TDF in North America. This latitudinal gradient is accompanied by gradual changes in climatic conditions; therefore, differences in functional traits from site to site along this gradient should be expected. Unfortunately, complete quantitative information for these sites is still not available. However, based on the regression model developed for the Chamela forest by Martínez-Yrízar et al. (1992), the aboveground biomass of trees in Alamos and San Javier varies from 45 to 118 Mg ha^{-1} (table 2.1). The higher biomass values suggest that some forests in the Alamos region could represent another type of dry forest, such as *bosque tropical subcaducifolio*.

Recent work in the southern Sonoran Desert in Mexico has shown that total aboveground phytomass on the northern slopes with tropical thornscrub and in the arroyos with xeroriparian vegetation is 10.5 and 24.4 Mg ha^{-1} y^{-1}, respectively. These values are closer to the lower limits reported for tropical dry forests and confirm the subtle boundaries separating true Sonoran Desert (6.5 Mg ha^{-1} y^{-1}), thornscrub, and TDF (Búrquez et al. 1992b, 1999). More striking are the calculations of NPP inferred from litter production (Angelina Martínez-Yrízar and Alberto Búrquez, unpubl. data). Preliminary results from a 4-year study in the Sonoran Desert show a total litterfall of 1.6 and 3.4 Mg ha^{-1} y^{-1} in the thornscrub and arroyo vegetation, respectively. The mean value for xeroriparian ecosystems in the Sonoran Desert is as high as the mean value for the core TDF of Chamela.

A summary of the major attributes of selected sites with tropical vegetation along the Pacific coast is shown in figure 2.2. The values of canopy height and aboveground phytomass are very similar between sites classified

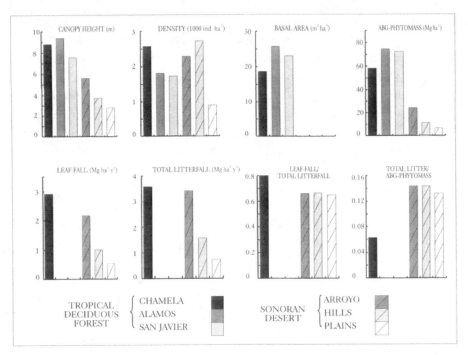

FIGURE 2.2. Major attributes of selected sites with tropical vegetation along the Pacific coast.

as TDF and decrease along an evapotranspiration gradient. However, if one assesses the quotients normalizing some functional traits of the ecosystem, such as the ratio of leaf-fall to total litterfall or the ratio of total litterfall to aboveground biomass, comparable results emerge. These ratios show similarities between the most xerophytic communities in the Sonoran Desert and the more mesic tropical deciduous forests (65–80% of the litterfall are leaves). They also show that desert systems are much more productive per unit of standing biomass. The annual litterfall in the desert represents about 14% of the standing aboveground biomass, twice the value of deciduous forests.

Conservation

Until recently the efforts to conserve areas with TDF were little rewarded. Before the late 1970s, there was not a single area along the Pacific coast of Mexico with an extensive section of protected forest. The first sizable reserve (1,600 ha) with TDF was acquired in this region by the Universidad Nacional Autónoma de México through a private donation, thus forming the core of the present research field station of Chamela, Jalisco. This area was later en-larged, also through a major private donation, to include a better represen-tation, and the designation changed from an academically oriented research facility to a Mexican biosphere reserve with 13,142 ha (Chamela-Cuixmala Biosphere Reserve; INE/CONABIO 1995; fig. 2.1). At about the same time, the reserves of Sierra La Laguna, Baja California Sur, and Sierra de Alamos–Río Cuchujaqui, Sonora, were finally decreed (Diario Oficial de la Federación 1996; INE/CONABIO 1995). These last two reserves protect the largest tracts of TDF in Mexico. The only other reserve protecting a sizable area of forest on the Pacific slope is located in Huautla, Morelos, in the Río Balsas Basin (31,314 ha; Morelos State Decree of March 31, 1993; fig. 2.1). Some reserves designed to protect other biomes include small TDF areas. Among these, Sierra de Manantlán in Jalisco, Sierra del Abra in San Luis Potosí, and Ría Lagartos in Yucatán are prominent (INE/CONABIO 1995). The tardiness and difficulties in the establishment of major reserves have left large unprotected areas in the Mexican states of Sinaloa, Nayarit, Guerrero, Michoacán, Oa-xaca, and Chiapas that have TDF as one of the predominant vegetation types.

Ecological studies on this ecosystem should be encouraged, not only on its basic biological features, but also on its functional aspects. The few studies already underway are helping us to identify sites within the Neotropical dry forest that should be given conservation priority and to define more con-servative management systems. The evaluation of diversity and endemisms, agroforesty techniques, multispecies forest plantations, and ecotourism will advance these goals.

NOTE

Several protected areas that include TDF were decreed after this chapter was submitted for publication. Especially important is the biosphere reserve Tehuacán-Cuicatlán, which pro-tects more than 150,000 ha in the Tehuacán Valley between Puebla and Oaxaca (fig. 2.1).

ACKNOWLEDGMENTS

We thank Professor Paul S. Martin, who revealed to us some of the Sierra Madre treasures: the Sonoran tropical deciduous forests. He has always been a source of inspiration. P. Cen-teno, R. Díaz, R. Esteban, H. Miranda, S. Núñez, M. A. Quintana, M. Villegas, and other students at the Instituto de Ecología, Universidad Nacional Autónoma de México, contrib-uted to data collection and analyses. Data from Sierra de Alamos were collected with the aid of students in field courses organized by the Tucson Audubon Society. This paper was partly supported by a Dirección General de Asuntos del Personal Académico–UNAM grant, PAPIIT-212894.

REFERENCES

Arriaga, L., and J. L. León de la Luz. 1989. The Mexican deciduous forest of Baja California Sur: A floristic and structural approach. *Vegetatio* 84:45–52.

Aznar, L. A. 1993. Contenido calórico de diferentes componentes estructurales de la vegetación de una selva baja caducifolia en Chamela, Jalisco, México. Tesis de Licenciatura. Facultad de Ciencias, Universidad Nacional Autónoma de México, México, D.F.

Bañuelos, N., and A. Búrquez. 1996. Las plantas: Una estrategia de salud en la medicina doméstica mayo. *Estudios Sociales* 12:163–189.

Bullock, S.H.B. 1988. Rasgos del ambiente físico y biológico de Chamela, Jalisco, México. *Folia Entomológica Mexicana* 77:5–17.

Bullock, S.H.B., and J. A. Solís-Magallanes. 1990. Phenology of canopy trees of a tropical deciduous forest in Mexico. *Biotropica* 21:22–35.

Búrquez, A. 1997. Distributional limits of Euglossine and Meliponine bees (Hymenoptera: Apidae) in northwestern Mexico. *Pan-Pacific Entomologist* 73:137–140.

Búrquez, A., A. Martínez-Yrízar, R. S. Felger, and D. Yetman. 1999. Vegetation and habitat diversity at the southern edge of the Sonoran Desert. In *Ecology of Sonoran Desert plants and plant communities*, ed. R. H. Robichaux, 36–67. Tucson: University of Arizona Press.

Búrquez, A., A. Martínez-Yrízar, and P. S. Martin. 1992a. From the high Sierra Madre to the coast: Changes in vegetation along Highway 16, Maycoba-Hermosillo. In *Geology and mineral resources of northern Sierra Madre Occidental, Mexico*, eds. K. F. Clark, J. Roldán-Quintana, and R. H. Schmidt, 239–252. El Paso: El Paso Geological Society.

Búrquez, A., A. Martínez-Yrízar, S. Núñez, T. Quintero, and A. Aparicio. 1992b. Above-ground phytomass of a Sonoran Desert community. *American Journal of Botany* (Supplement-Abstracts) 79:186.

Bye, R. 1995. Ethnobotany of the Mexican tropical dry forests. In *Seasonally dry tropical forests*, eds. S. H. Bullock, H. A. Mooney, and E. Medina, 423–438. Cambridge: Cambridge University Press.

Castellanos, J., J. M. Maass, and J. Kummerow. 1991. Root biomass of a tropical deciduous forest. *Plant and Soil* 24:270–274.

Cox, G. W., and M. D. Atkins. 1979. *Agricultural ecology*. San Francisco: W. H. Freeman.

Dávila, P., J. L. Villaseñor, R. Medina, A. Ramírez, A. Salinas, J. Sánchez-Ken, and P. Tenorio. 1993. *Flora del Valle de Tehuacán-Cuicatlán*. Listados Florísticos de México X. México, D.F.: Instituto de Biología, Universidad Nacional Autónoma de México.

De Ita-Martínez, C. 1983. Patrones de producción agrícola en un ecosistema tropical estacional en la costa de Jalisco. Tesis de Licenciatura. Facultad de Ciencias, Universidad Nacional Autónoma de México, México, D.F.

Diario Oficial de la Federación. 1996. *Area de Protección de flora y fauna silvestres y acuáticas, Sierra de Alamos-Río Cuchujaqui*. México, D.F.: Tomo DXIV, no. 15, sección 1:14–19.

Felger, R. S. 1999. The flora of Cañón del Nacapule: A desert-bounded tropical canyon near Guaymas, Sonora, Mexico. *Proceedings of the San Diego Society of Natural History* 35:1–42.

Filip, V., R. Dirzo, J. M. Maass, and J. Sarukhán. 1995. Within- and among-year variation in the levels of herbivory on the foliage of trees from a Mexican tropical deciduous forest. *Biotropica* 27:78–86.

Flores-Villela, O., and P. Geréz. 1988. *Conservación en México: Síntesis sobre vertebrados terrestres, vegetación y uso del suelo.* México, D.F.: INIREB/CI.

Galicia, L. 1992. Influencia de la variabilidad de la forma de la pendiente en las propiedades físicas del suelo y su capacidad de retención de agua, en una cuenca tropical estacional. Tesis de Licenciatura. Facultad de Filosofía y Letras, Universidad Nacional Autónoma de México, México, D.F.

García, E. 1973. *Modificaciones al sistema de clasificación climática de Köeppen, para adaptarto a las condiciones de la Republica Mexicana.* México, D.F.: Instituto de Geografía, Universidad Nacional Autónoma de México.

García-Oliva, F., E. Ezcurra, and L. Galicia. 1991. Pattern of rainfall distribution in the Central Pacific Coast of Mexico. *Geografiska Annaler* 73(A):3–4.

García-Oliva, F., and M. Maass. 1998. Efecto de la transformación de la selva a pradera sobre la dinámica de los nutrientes en un ecosistema tropical estacional en México. *Boletín de la Sociedad Botánica de México* 62:39–48.

Gentry, A. H. 1995. Diversity and floristic composition of neotropical dry forests. In *Seasonally dry tropical forests*, eds. S. H. Bullock, H. A. Mooney, and E. Medina, 146–194. Cambridge: Cambridge University Press.

Gentry, H. S. 1942. *Río Mayo plants: A study of the flora and vegetation of the valley of the Río Mayo, Sonora.* Publication No. 527. Washington, D.C.: Carnegie Institution of Washington.

———. 1946. Sierra Tacuichamona: A Sinaloa plant locale. *Bulletin of the Torrey Botanical Club* 73:356–362.

Goldberg, D. E. 1982. The distribution of evergreen and deciduous trees relative to soil type: An example from the Sierra Madre, Mexico, and a general model. *Ecology* 63:942–951.

———. 1985. Effects of soil pH, competition, and seed predation on the distribution of two tree species. *Ecology* 66:503–511.

Gual, M. 1995. *Cañón del Zopilote (Area Venta Vieja).* Estudios Florísticos de Guerrero No. 6, eds. N. Diego-Pérez and R. M. Fonseca. México, D.F.: Facultad de Ciencias, Universidad Nacional Autónoma de México.

INE/CONABIO. 1995. *Reservas de la Biosfera y otras areas naturales protegidas de México.* México, D.F.: Instituto Nacional de Ecología, Comisión Nacional para el Conocimiento y Uso de la Biodiversidad.

INEGI. 1998. *Atlas nacional del medio físico.* México, D.F.: Instituto Nacional de Geografía, Estadística e Informática, Aguascalientes.

Janzen, D. H. 1988. Tropical dry forests: The most endangered major tropical ecosystem. In *Biodiversity*, ed. E. O. Wilson, 130–137. Washington, D.C.: National Academy of Sciences Press.

Jaramillo-Luque, V. 1992. El fuego y la biogeoquímica en un ecosistema tropical estacional. *Ciencia* 43:41–43.

Jaramillo-Luque, V., and R. L. Sanford. 1995. Nutrient cycling in tropical deciduous forests. In *Seasonally dry tropical forests*, eds. S. H. Bullock, H. A. Mooney, and E. Medina, 346–361. Cambridge: Cambridge University Press.

Jaramillo-Villalobos, V. 1994. *Revegetación y reforestación de las áreas ganaderas en las zonas tropicales de México*. México, D.F.: COTECOCA, Secretaría de Agricultura y Recursos Hidráulicos.

Johnson, M. B. 1992. The genus *Bursera* (Burseraceae) in Sonora, Mexico, and Arizona, U.S.A. *Desert Plants* 10:126–144.

Joyal, E. 1996. The use of *Sabal uresana* (Arecaceae) and other palms in Sonora, Mexico. *Economic Botany* 50:429–445.

Kauffman, B. J., V. Jaramillo-Luque, R. Dirzo, and R. L. Sanford Jr. 1994. *Biomass and carbon dynamics of tropical forests of Mexico*. Corvallis, Oreg.: Environmental Research Lab, Oregon State University, CR 821860 Progress Report.

Kohlman, B., and S. Sánchez-Colón. 1984. *Estudio areográfico del género* Bursera *en México*. México, D.F.: Instituto de Ecología, A.C.

Krizman, R. D. 1972. Environment and season in a tropical deciduous forest in northwestern Mexico. Ph.D. dissertation, University of Arizona, Tucson.

Kummerow, J., J. Castellanos, M. Maass, and A. Larigauderie. 1990. Production of fine roots and the seasonality of their growth in a Mexican deciduous dry forest. *Vegetatio* 90:73–80.

Lal, R. 1990. *Soil erosion in the tropics: Principles and management*. New York: McGraw-Hill.

Lott, E. J. 1993. *Annotated checklist of the vascular flora of the Chamela Bay region, Jalisco, Mexico*. Occasional Papers of the California Academy of Sciences No. 148.

Lott, E. J., S. H. Bullock, and J. A. Solís-Magallanes. 1987. Floristic diversity and structure of upland and arroyo forests of coastal Jalisco. *Biotropica* 19:228–235.

Lumholtz, C. 1902. *Unknown Mexico: A record of five years exploration among the tribes of the western Sierra Madre, in the Tierra Caliente of Tepic and Jalisco, and among the Tarascos of Michoacán*. New York: C. Scribners & Sons.

Maass, J. M. 1995. Conversion of tropical dry forest to pasture and agriculture. In *Seasonally dry tropical forests*, eds. S. H. Bullock, H. A. Mooney, and E. Medina, 399–422. Cambridge: Cambridge University Press.

Maass, J. M., C. Jordan, and J. Sarukhán. 1988. Soil erosion and nutrient losses in seasonal tropical agroecosystems under various management techniques. *Journal of Applied Ecology* 25:595–607.

Maass, J. M., J. M. Vose, W. T. Swank, and A. Martínez-Yrízar. 1995. Seasonal changes of leaf area index (LAI) in a tropical deciduous forest in west Mexico. *Forest Ecology and Management* 74:171–180.

Martin, P. S., D. A. Yetman, M. E. Fishbein, P. D. Jenkins, T. R. Van Devender, and R. Wilson, eds. 1998. *Gentry's Río Mayo plants: The tropical deciduous forest and environs of northwest Mexico*. Tucson: University of Arizona Press.

Martínez-Yrízar, A. 1980. Tasas de descomposición de materia orgánica foliar de especies arbóreas de selvas en clima estacional. Tesis de Licenciatura. Facultad de Ciencias, Universidad Nacional Autónoma de México, México, D.F.

———. 1995. Biomass distribution and primary productivity of tropical dry forests. In *Seasonally dry tropical forests*, eds. S. H. Bullock, H. A. Mooney, and E. Medina, 326–345. Cambridge: Cambridge University Press.

Martínez-Yrízar, A., M. Maass, L. A. Pérez-Jiménez, and J. Sarukhán. 1996. Net primary productivity of a tropical deciduous forest ecosystem in western Mexico. *Journal of Tropical Ecology* 12:169–175.

Martínez-Yrízar, A., and J. Sarukhán. 1990. Litterfall patterns in a tropical deciduous forest in Mexico over a five-year period. *Journal of Tropical Ecology* 6:433–444.

———. 1993. Cambios estacionales de mantillo en el suelo de un bosque tropical caducifolio y uno subcaducifolio en Chamela, Jalisco, México. *Acta Botánica Mexicana* 21:1–6.

Martínez-Yrízar, A., J. Sarukhán, A. Pérez-Jiménez, E. Rincón, J. M. Maass, A. Solís-Magallanes, and L. Cervantes. 1992. Above-ground phytomass of a tropical deciduous forest on the coast of Jalisco, Mexico. *Journal of Tropical Ecology* 8:87–96.

Miranda, F. 1941. Estudios sobre la vegetación de México. I. La vegetación de los cerros al sur de la meseta de Anáhuac—el cuajiotal. *Anales del Instituto de Biología* 12:569–614.

———. 1947. Estudios sobre la vegetación de México. V. Rasgos de la vegetación en la Cuenca del Río Balsas. *Revista de la Sociedad Mexicana de Historia Natural* 8:95–114.

———. 1952. *La vegetación de Chiapas*. Tuxtla Gutiérrez: Ediciones del Gobierno del Estado de Chiapas.

Miranda, F., and E. Hernández-X. 1963. Los tipos de vegetación de México y su clasificación. *Boletín de la Sociedad Botánica de México* 28:29–179.

Murphy, P. G., and A. E. Lugo. 1986. Ecology of tropical dry forest. *Annual Review of Ecology and Systematics* 17:67–88.

———. 1995. Dry forests of Central America and the Caribbean. In *Seasonally dry tropical forests*, eds. S. H. Bullock, H. A. Mooney, and E. Medina, 9–34. Cambridge: Cambridge University Press.

Ochoterena, I. 1923. Las regiones geográfico-botánicas de México. *Revista de la Escuela Nacional Preparatoria* 1:261–331.

Patiño, C. 1990. Variación espacial y temporal de la capa de hojarasca (mantillo) en una selva baja caducifolia en Chamela, Jal., México. Tesis de Licenciatura. Facultad de Ciencias, Universidad Nacional Autónoma de México, México, D.F.

Peralta, S. 1995. *Cañón del Zopilote, Area Papalotepec*. Estudios Florísticos de Guerrero No. 5, eds. N. Diego-Pérez and R. M. Fonseca. México, D.F.: Facultad de Ciencias, Universidad Nacional Autónoma de México.

Rzedowski, J. 1978. *La vegetación de México*. México, D.F.: Limusa-Wiley.

Rzedowski, J., and R. McVaugh. 1966. La vegetación de Nueva Galicia. *Contributions from the University of Michigan Herbarium* 9:1–123.

Sarukhán, J., and M. Maass. 1990. Bases conceptuales para un manejo sostenido de los ecosistema: El sistema de cuencas hidrológicas. In *Medio ambiente y desarrollo en México*, vol. 1, ed. E. Leff, 81–114. México, D.F.: UNAM (CIIH)-Porrúa.

Toledo, V. M. 1988. La diversidad biológica de México. *Ciencia y Desarrollo* 8:7–16.

Vargas, A., and A. Pérez. 1996. *Cerro Chiletépetl y Alrededores (Cuenca del Balsas)*. Estudios

Florísticos de Guerrero No. 7, eds. N. Diego-Pérez and R. M. Fonseca. México, D.F.: Facultad de Ciencias, Universidad Nacional Autónoma de México.

Whittaker, R. H. 1975. *Communities and ecosystems*. New York: Macmillan.

Wiseman, F. M. 1980. The edge of the tropics: The transition from tropical to subtropical ecosystems in Sonora, Mexico. *Geoscience and Man* 21:141–156.

Yetman, D. A., and A. Búrquez. 1994. Buffelgrass: Sonoran Desert nightmare. *Arizona Riparian Council Newsletter* 3:8–10.

———. 1996. A tale of two species: Speculation on the introduction of *Pachycereus pringlei* in the Sierra Libre, Sonora, Mexico, by *Homo sapiens*. *Desert Plants* 12:23–31.

Yetman, D. A., T. R. Van Devender, P. Jenkins, and M. Fishbein. 1995. The Río Mayo: A history of studies. *Journal of the Southwest* 37:294–345.

3

VEGETATION, FLORA, AND SEASONS OF THE RÍO CUCHUJAQUI, A TROPICAL DECIDUOUS FOREST NEAR ALAMOS, SONORA

Thomas R. Van Devender, Andrew C. Sanders,
Rebecca K. Wilson, and Stephanie A. Meyer

The pueblo of Alamos is the dowager queen of southern Sonora, immersed in memories of an elegant and rich past, steeped in its cultural heritage. Between 1533 and 1540, the Spanish conquistadores Diego de Guzmán and Vásquez de Coronado and the famous traveler Alvaro Nuñez Cabeza de Vaca passed by the site of Alamos and the distinctive Los Frailes peaks of the Sierra de Alamos on Mayo Indian trails (French 1962). In 1630 Jesuit missionaries built an adobe church on the spot where the Iglesia de Alamos now stands, essentially founding Real de los Frailes, New Spain, later to become Alamos, Sonora, Mexico. After the 1683 discovery of fabulously rich silver and gold deposits, Alamos thrived as a mining and religious center. Many of the expeditions that established missions in Pimería Alta (northern Sonora and southern Arizona) and in such distant places as Los Angeles, San Francisco, and Monterrey were funded with silver from the Alamos mines.

About 1940, wealthy Americans "discovered" the town, which was mostly in ruins, 450 km south of the Arizona border and were smitten with its Spanish colonial ambience, the beauty of the tropical forest, the scenic Sierra de Alamos, and the tranquil pace of life. During the last half century, the influx of "silver" from the United States and Canada has resulted in the resurgence of Alamos. Renovating the old *mansiones* with local materials, labor, and skills is an important local industry. The result is a unique mosaic of historical buildings, colonial architecture, cobblestone streets, cultural traditions of religious celebrations, dances, the Sunday night promenade, a historical museum, a library, a classical music festival, and a mixed Mexican and American population that has lived together for several generations.

But for the biologist, the charm of Alamos is only the beginning. The variety of habitats in northwestern Mexico is breathtaking. In only 800 km, a traveler can see temperate mixed-conifer, ponderosa pine, and Madrean pine-oak forests in the Santa Catalina Mountains above Tucson; oak woodland and desert grassland in southern Arizona and northern Sonora; Sonoran Desert and foothills thornscrub in central Sonora; coastal thornscrub and mangrove swamps along the coast of the Gulf of California; and tropical deciduous forest (TDF) in southern Sonora. The changes in species, vegetation, and climate both from north to south and from sea level to the inland mountains are dramatic. For the North American traveler, Alamos is the gateway to the New World tropics, the northernmost opportunity to experience the tropical dry season and the overwhelming diversity of the lush summer monsoon forest: rampant vines and lianas; epiphytes; exotic animals, including boa constrictors *(Boa constrictor)*, brown vine snakes *(Oxybelis aeneus)*, Amazon parrots *(Amazona albifrons, A. finschi)*, and black-throated magpie jays *(Calocitta formosa colliei)*; and other tropical specialties.

On the western side of the continent, the northern limits of the New World tropics are in central Sonora, with a relatively abrupt transition between 28° and 32°N latitude, 8–10° farther north than on the east coast of Mexico. Although Gentry's (1942) *Río Mayo Plants* was one of the earliest studies of TDF (as *short tree forest*), the tropical communities of northwestern Mexico have been neglected, probably because of the ease of air travel from the United States to Costa Rica and other low-latitude tropical areas and the distance of more than 1,300 km between Alamos, Sonora, and Mexico City. The only important Mexican study area in *Seasonally Dry Tropical Forests* (Bullock et al. 1995), a summary of dry forest studies from many areas in the New World, is the Estación de Biología Chamela (19°32'N) in Jalisco.

In this paper we discuss the vegetation structure, seasons, flora, and floristic affinities of TDF in the Alamos area, focusing on the Río Cuchujaqui, a stream originating in the Sierra Madre Occidental of western Chihuahua.

In March 1890, Edward J. Palmer, an English professional plant collector, ventured from Guaymas, a port on the Gulf of California, to Alamos, then a silver-mining town of about 10,000 inhabitants. He spent two weeks collecting plants at the peak of the spring dry season. The tropical flora proved to be of such interest that he returned for another two weeks in September, this time via steamer to Agiabampo, then the seaport of Alamos, to collect at the peak of the summer rainy season. Palmer's plant collections from the Alamos area, which were reported by Rose in 1891, included 43 new species—an amazing 18.9% of his 227 identifiable (36 were not) collections! Unfortunately, quite a few of Palmer's new species from the Alamos area have not survived the taxonomic turmoil of this century; many were merged into widespread, variable tropical species that were described earlier.

Most of Palmer's collections were from the immediate vicinity of Alamos or the nearby Sierra de Alamos. The labels for a few species *(Guardiola platyphylla, Ipomopsis sonorae, Iresine celosia, Perityle microglossa,* and *Solanum erianthum)*, which read "from river banks," "along river," or "from sandy river bottoms," were likely from the Río Cuchujaqui, perhaps at the modern Güirocoba crossing about 12 km south-southeast of Alamos. His collection number 698 from "Many plants growing together on rocky ledges, Alamos" became the type specimen of *Bouteloua alamosana* (Alamos grama grass). This proved to be an elusive species that was not re-collected near Alamos until the present study. It is a locally abundant, summer annual restricted to depressions on top of bluffs at several places along the Río Cuchujaqui; Palmer's type collection could have been from this area as well.

Between the fall of 1933 and November 1939, Howard Scott Gentry made 3,200 collections of 1,276 species in southern Sonora and west-central Chihuahua. His *Río Mayo Plants,* published in 1942, is a remarkable blend of the adventures of a young man in search of botanical treasures in the remote, mysterious Sierra Madre Occidental, portraits of fascinating and little-known landscapes and native peoples, and a scholarly work of great depth with poetic descriptions rarely found in scientific works. The Alamos area was not studied as extensively as other areas in the Río Mayo region because it was mostly in the Río Fuerte drainage to the south. Surveys of Gentry's specimens in the University of Arizona Herbarium yielded only a few specimens collected in 1933 and 1936 from "Arroyo Cuchujaqui" *(Aphanosperma sinaloensis, Iresine interrupta, Mecardonia vandellioides, Melochia tomentella, Pseudognaphalium leucocephalum*, and *Solanum ferrugineum*).

Study Area

The Río Cuchujaqui is the northernmost tributary of the Río Fuerte and drains most of southeastern Sonora that lies south of the Río Mayo. From its headwaters near Chínipas in the Sierra Madre Occidental of southwestern Chihuahua, the river flows west into southeastern Sonora, then turns south (as the Río de Alamos on some maps) into Sinaloa. Historically, the channel joined the Río Fuerte in Sinaloa. With the construction of the Presa Josefa Ortíz Domínguez, the water was diverted away from the Río Fuerte to the agricultural areas near Agiabampo, Sonora (Alejandro Varela, pers. comm. 1994).

The study area extends from the area upstream of the crossing at Rancho El Guayabo, about 35 km west of the Chihuahua border, west and south to the La Ranchería crossing, a distance of 42 km along the river and 25 km by air (fig. 3.1). Elevation drops from almost 400 m on the slopes above El Guayabo to 220 m at La Ranchería. Plants were surveyed at nine localities along the Río Cuchujaqui, ranging from 14 km east-southeast to 22.5 km south-southeast of Alamos. Plants were also surveyed along Arroyos Alamos and El Mentidero above their junction with the river. We believe that the total surveyed area of about 21 km² is representative of the 46 km² in the entire study area.

For much of its length, the Río Cuchujaqui has cut into the conglomeritic mudflow deposits of the Báucarit Formation, resulting in cliffs and benches on one or both sides of the river. Similar deposits in many areas in eastern and southern Sonora have been assigned ages ranging from late Oligocene to late Miocene (ca. 25 to 5 million years ago; Robert Scarborough, pers. comm. 1999). The weathered surfaces are coated with lichens, which give a dark gray color to the exposed rock. The cliffs are very steep in many places, supporting either no plants or very interesting species that are rarely seen in other habitats; *Russelia sonorensis*, a small shrub with brilliant red flowers, is one such species. In many areas, steep slopes, often covered with *Agave vilmoriniana* (octopus agave, amole), are restricted to one side of the river. In other areas, the cliffs form deep canyons or *cajones* with narrow, rocky stream bottoms.

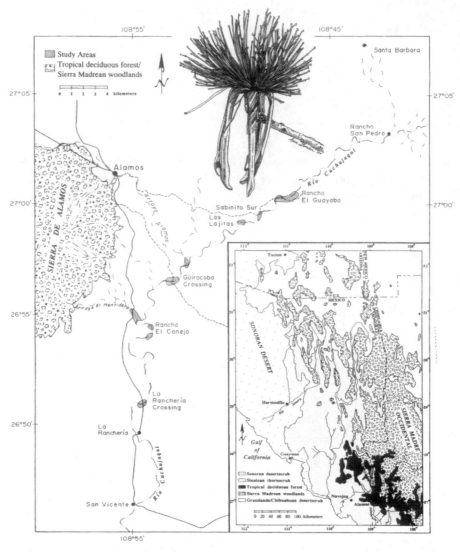

FIGURE 3.1. Map of the Alamos area of southern Sonora. Study areas along the Río Cuchujaqui are stippled. Inset vegetation map modified from Brown and Lowe (1980). The flower, drawn by Anne E. Gondor, is *Pseudobombax palmeri*.

Water is permanently available along the river; this is the only zone in the area that can support significant numbers of evergreen trees. During the dry season, the Río Cuchujaqui is a ribbon of green vegetation twisting through the gray-brown hills covered with leafless TDF. The rocky slopes above the river support drought deciduous and succulent species as well as annual and perennial herbs. Under the permanently moist conditions along the river, a vegetation approaching tropical semideciduous forest could develop locally, except that frequent and often severe floods reduce the area of the river bottom that can be occupied by evergreens. Only plants along the margins of the canyon can expect long lives. The evergreen best adapted to the severe flooding is the massive *Taxodium distichum* var. *mexicanum* (Mexican bald cypress, sabino), which is the most conspicuous tree in frequently flooded areas and typically the only species in the middle of the channel. But even this giant is occasionally knocked over during larger floods and appears to have some trouble in establishment; seedlings and saplings are not common.

At Rancho El Guayabo, 2.6 km northeast of Sabinito Sur, 14 km east of Alamos (27°00'05"–01'30"N, 108°46'15"–47'45"w; 340–360 m), the road to Rancho San Pedro crosses the river twice. The riverbed is mostly sand, gravel, and boulders with only a few deeper bedrock pools in canyons. Riparian gallery forests are well developed. Just above the upper crossing, both banks are lined by tall *Taxodium distichum*, whose branches hang over a quietly flowing stretch of water. In the dry season, the river is about 6 m wide and 0.5 m deep and flows over water-tumbled stones covered with the mosslike *Oserya* cf. *coulteri* (rock water moss). The interlaced roots of the trees lock the rocks together, stabilizing the banks to such a degree that the scene has not changed noticeably in 20 years despite annual flooding.

Hymenocallis sonorensis (spider lily, lirio de los arroyos), with its elegant white flowers and dark, broad, flat leaves in dense meter-tall clumps, is common in the stream. *Salix taxifolia* (yewleaf willow, sauce), unknown elsewhere in southern Sonora, is a locally common shrub at the edge of the stream. The sandy, needle-littered soil underneath the sabinos is a rich habitat for shade-loving herbs, such as *Desmodium scorpiurus* (sabino tick clover, frijol de godorniz [= codorniz]), a prostrate legume with a rose-colored flower; *Habenaria quinqueseta* (spider orchid), a terrestrial orchid with a large, elegant white flower; and *Xanthosoma hoffmannii* (polongo). Unfortunately for the unwary visitor, the shiny-leaved vines found on many of the sabino trunks are *Rhus radicans* (poison ivy, hiedra).

At El Guayabo, TDF extends from 420 m on the slopes down into the canyon to meet the riparian gallery forest. Secluded forests in tributary canyons are protected from the onslaught of the summer floods and often harbor tropical plants that are rare or absent in other areas. The most prominent and typically the largest trees in such canyons are the evergreen figs, especially *Ficus cotinifolia* (strangler fig, nacapuli), *F. pertusa* (camuchín), and

F. petiolaris (rock fig, tescalama). A side canyon above the crossing supports a stand of *Platymiscium trifoliolatum* (tampicerán), a 12-m tropical tree with bright yellow flowers on leafless stems in March. *Brahea aculeata* (palmilla), a handsome small palm, and *Piper jaliscanum* (cocolmeca), a tropical shrub, grow in the same canyon. The giant herb *Eupatorium quadrangulare* (cocolmeca grande) seems to occur only in these shaded alcoves, where it is typically common. Succulents are notably scarce in these shady forests but may be very common on adjacent cliffs and dry slopes.

Sabinito Sur is a picturesque village of palm-thatched houses 12.5 km east-southeast of Alamos (26°59'45"N, 108°48'30"W; 320 m). The Las Lajitas crossing is 1.8 km downstream from Sabinito Sur and 11 km east-southeast of Alamos (26°59'05"N, 108°49'45"W; 320 m). Just above the crossing, there is a long, deep pool bordered by steep cliffs. On top of the cliffs, a small pond in a bedrock depression supports aquatic plants, including the blue-flowered *Heteranthera limosa* (mud plantain), the white-flowered *Limnocharis flava* (flowering rush), and *Najas guadalupensis* (common water nymph).

The Arroyo Alamos (also called Arroyo Aduana and Arroyo Las Cabras) enters the Río Cuchujaqui 9.3 km southeast of Alamos (26°57'30"N, 108°52'15"W; 280 m). Here the river and the Arroyo Alamos form narrow canyons *(cajones)* with strings of long, deep bedrock pools *(tinajas)* punctuated by coarse boulder-strewn bars. The riparian vegetation is sparse, with only a few massive *Taxodium distichum* able to survive the intense annual floods.

The Güirocoba crossing is near Rancho El Cuchujaqui, 12.3 km south-southeast of Alamos (26°56'15"N, 108°53'W; 260 m). A bridge over the river completed in 1993 to facilitate travel between Alamos and Güirocoba was washed away by floods in September 1995, only to be promptly rebuilt. The river alternates between the cajón-tinaja environment and open stretches with the river channel on shallow gravel beds. Where the water is permanent but the channel is wide enough to dampen the annual floodwater, sabino-sauce (*Salix bonplandiana* and *S. gooddingii*) gallery forest is very well developed. *Chloracantha spinosa* var. *jaliscensis* (water aster) grows only in the water, whereas *Ambrosia ambrosioides* (canyon ragweed, chicura), *Baccharis salicifolia* (seep willow, batamote), *Hymenoclea monogyra* (cheesebush, romerillo), and *Montanoa rosei* (batayaqui) are common shrubs in sand, gravel, or cobble in frequently flooded areas.

The only specimens of *Populus mexicana* subsp. *dimorpha* (Mexican cottonwood, álamo) in the study area were planted by Ismael Acosta decades ago and in 1993 near the Güirocoba crossing. Ironically, the massive tree and young saplings were washed away by the floods of Hurricane Ismael in September 1995. These massive trees with both heart-shaped cottonwood and linear willowlike leaves are more typically found on floodplains of the Río Mayo and the Río Yaqui and in the town of Alamos. Apparently, they are naturally absent from the Río Cuchujaqui study area because they are more vulnerable to the floods than sabinos or sauces.

Several interesting habitats occur on top of the bluffs along the river. The volcanic ash bedrock is exposed on broad, flat benches with shallow or no soil. Large areas are relatively barren or support very open TDF. *Jatropha malacophylla* (sangrengado), a tall shrub with smooth, reddish brown bark, succulent stems, and sycamore-like leaves, and *Malpighia emarginata* (granadilla), a small tree with mottled bark and pink flowers, are especially common on these ledges. Many interesting and unusual herbs, such as *Bouteloua alamosana, B. quiriegoensis* (Quiriego grama), and *Fimbristylis dichotoma*, live in depressions in the bedrock where water stands and thin organic soils develop.

An enormous *Ficus pertusa* in a protected side canyon is the largest tree in the study area. The "trunk" is actually a complex of several fused trunks and massive aerial roots that defies easy definition or measurement. Both the height and greatest crown width are perhaps 25 to 30 m.

The Arroyo El Mentidero drainage is a series of steep canyons on the eastern slopes of the Sierra de Alamos that coalesce and join the Río Cuchujaqui 12.3 km south of Alamos (26°54'45"N, 108°55'05"W; 240 m). In the last few kilometers above the river, the arroyo has cliffs on each side, a gravel bed, scattered large boulders, and a seasonal stream. The river canyon near the junction is broad with steep cliffs on the west bank and gentler slopes across the river. *Cephalanthus occidentalis* (button bush, mimbro), a tall shrub with balls of fragrant white flowers, is common at the edge of the stream. The pale trunks of occasional *Albizia sinaloensis* (palo joso) are visible just above the sabino-sauce gallery forest along the river. The side canyons and arroyo margins are locally dominated by *Celtis iguanea* (tropical hackberry, cumbro) and *Pisonia capitata* (garambullo), both of which are thorny, scandent shrubs or woody vines that in places form impenetrable thickets along the river. Just below the junction with the Río Cuchujaqui, there is a large bend in the river at Rancho El Conejo at the 220-m elevation, 12.4 to 13.4 km south-southeast of Alamos (26°54'35"N, 108°55'05"W; 220 m).

The La Ranchería crossing is just north of the village of La Ranchería,

FIGURE 3.2. Vegetation profile from the coast of the Gulf of California through the Sierra de Alamos to the Río Cuchujaqui in southern Sonora.

22.5 km south-southeast of Alamos (26°51'N, 108°55'W; 200 m). The river is in a broad, open canyon with rocky slopes; a few bedrock cliffs overlook the east side. Silts and sands dominate the muddy streambed, which is dotted with gravel bars and bedrock pools. TDF is best developed on north-facing scarps above the river; elsewhere the forest is beginning to take on the appearance of, and share species with, the more xeric coastal thornscrub to the south and southwest.

By the time the Río Cuchujaqui reaches San Vicente 27 km south of Alamos (26°45'20"N, 108°55'W; 180 m), it is a broad, shallow stream in a muddy bed lined with *Prosopis glandulosa* (honey mesquite, mezquite), a variety of shrubs, and occasional *Albizia sinaloensis* and *Taxodium distichum*. Near Los Muertos 33 km south of Alamos (26°44'15"N, 108°54'30"W; 170 m), a few *Populus mexicana* and *Salix gooddingii* (Goodding willow, sauce) are found along the river channel. Areas away from the river and at lower elevations westward toward the coast of the Gulf of California support foothills and coastal thornscrub (fig. 3.2; Friedman 1996).

Methods

Except for 1980–1981, 1985–1987, and 1990, we made collections on the river in each year between 1975 and 1995. The objectives of the recent phase of the study were to inventory the entire vascular plant flora at each of nine areas along the Río Cuchujaqui in all seasons and to collect voucher specimens of them. Collections have been made in every month of the year with somewhat less coverage from May to August.

Data on a total of 2,300 numbered collections were available for this study. Most (93.3%) of the collections were made by us. Van Devender and Wilson made 1,505 collections on the river on 51 field days between 1976 and 1995. Sanders made 628 collections on 16 field days between 1975 and 1993. Meyer has visited the river regularly since she moved to Alamos in November 1989 and made 24 independent collections. Most (93.2%) of the collections are from five localities: the El Guayabo crossing (31.8%), the Güirocoba crossing (25.2%), the Arroyo El Mentidero junction (19.7%), the La Ranchería crossing (10.4%), and the Las Lajitas crossing (6.1%).

Data on additional collections were provided by Victor W. Steinmann (32), Richard S. Felger (20), Paul S. Martin (18), Greg Starr (17), Thomas F. Daniel and Mary Butterwick (16), Charles T. Mason (16), Marc A. Baker (9), Martha A. Burgess (9), Mark E. Fishbein (8), Robert Perrill (7), Howard S. Gentry (6), Julia V. Salmon (3), Richard D. Krizman (1), and Ceal Smith (1). Data on 1,658 plant records from our collections, University of Arizona Herbarium specimens, and field notes provided by various collectors were entered into a computer database.

At least 3,943 specimens (counting duplicates) from the study area are deposited in 32 herbaria. Primary sets of voucher specimens were deposited into herbaria at the University of Arizona (ARIZ), University of California at Riverside (UCR), Arizona State University, and the Universidad de Sonora. Additional collections were deposited at the University of Texas at Austin, California Academy of Sciences, Missouri Botanical Garden, New York Botanical Garden, and the various herbaria of collaborating specialists.

Specimens were identified using regional floras (Gentry 1942; Kearney and Peebles 1969; McVaugh 1983, 1984, 1987; Shreve and Wiggins 1964; Standley 1920–1926; Wiggins 1980), numerous monographs, and comparative material at ARIZ and UCR. Additionally, 51 specialists from 29 institutions (see Acknowledgments) provided determinations in their families.

Tropical Deciduous Forest

TDF grows in a continuous ribbon along the coast from southern Sonora south to Costa Rica, with an average width of only 50 km (Gentry 1995; Krizman 1972; fig. 3.3). Throughout this area, dry tropical forests have been heavily impacted by human activities. Pre-Columbian and historical cultural impacts on tropical forests are very difficult to evaluate. TDF in Sonora ap-

FIGURE 3.3. Distribution of tropical deciduous forest in North America. Modified from Reichenbacher et al. (1998).

pears to have been impacted much less than that in more heavily populated areas to the south. The most extensive areas of mature TDF today are likely in Sonora.

As mentioned earlier, Gentry (1942) called the TDF around Alamos *short tree forest*. Shelford (1963) noted that TDF could be divided into short tree communities from southern Sonora south to Chiapas and tall tree communities from there to Panama. Krizman (1972) provided an excellent literature review of the history of the forest names. Rzedowski (1978) and Rzedowski and McVaugh (1966) considered the inland forests of Sinaloa and southern Sonora as northwestern extensions of the *bosque tropical caducifolio* or *bosque tropical deciduo* of Nueva Galicia, a large area centered on the states of Jalisco, Colima, and Aguascalientes. Although the compositions of the forests of Nueva Galicia are quite varied, there are local areas dominated by *Lysiloma divaricatum* (as palo de arco rather than mauto)—for example, near Autlán, Jalisco (Rzedowski 1978)—that resemble Sonoran forests. TDF in the Chamela Bay region of Jalisco is dominated by leguminous trees in the genus *Lonchocarpus* (12 species; Lott 1993) rather than *Lysiloma divaricatum*. Brown and Lowe (1980), Brown et al. (1979), and Gentry (1982) defined the tropical forests of Sinaloa and southern Sonora as *Sinaloan deciduous forest*. To the south in Nayarit and Jalisco, Sinaloan deciduous forest merges with tropical semideciduous forest. Because *L. divaricatum*–dominated forests are not unique to Sinaloa and Sonora but also grow toward the south, the Sonoran forests are best termed *tropical deciduous forest* rather than *Sinaloan deciduous forest* (Alberto Búrquez M., pers. comm. 1997).

A general overview of New World TDF (Bullock et al. 1995) included Chamela as the only Mexican site. TDF contains a greater diversity of life-forms, both in structure and physiology, than other tropical ecosystems (Medina 1995). Where the rainy season is 4–5 months long, the forests typically have lots of CAM plants (cacti, Agavaceae), succulent woody plants, deciduous woody plants, and vines, including lianas. The forests at Chamela had a greater diversity of trees than those in Guanacaste, Costa Rica (Gentry 1995). There are more regionally endemic genera in the dry forests of western Mexico than in those of Central America (Gentry 1995).

The temporal and spatial distribution of water availability is perhaps the most important determinant of the characteristics of tropical vegetation, such as canopy height, forest structure, annual growth and basal area of trees, and phenology (Bullock 1995). Because a landscape patch of less than 100 km^2 is highly diverse, forest composition and structure are also extremely variable within an area that pollinators and seed dispersers use for daily foraging or short-term movements. The aboveground phytomass is related to rainfall and successional status, although 9–50% of the total phytomass is allocated to roots (Cuevas 1995; Martínez-Y. 1995).

Two papers provide a great deal of information about the structure and composition of TDF near Alamos. In 1972 Richard D. Krizman completed a doctoral dissertation at the University of Arizona entitled "Environment and Season in a Tropical Deciduous Forest in Northwestern Mexico." His study area was just west of Minas Nuevas on the north flank of the Sierra de Alamos. The second description of the forest was written by Howard Scott Gentry himself in 1982, 40 years after he published *Río Mayo Plants!* The community is a heterogeneous, drought-deciduous forest with a variable but unbroken canopy 10–15 m above the ground. Around Alamos, this forest is mostly found at an elevation of 300–800 m in canyons or up to 1,000 m on rocky mountainsides. Here, as in Nueva Galicia (Rzedowski and McVaugh 1966), TDF is best developed on shallow rocky soils on mountain slopes rather than on deeper alluvial soils in broader valleys or on the coastal plain. Between 200 and 300 m, more xeric species are present in an ecotone between

TDF and thornscrub. Notable exceptions are found just east of Navojoa, where foothills thornscrub reaches nearly 400 m on the basaltic slopes of Cerro Prieto, and at Yocogigua on the south side of the Sierra de Alamos, where TDF is moderately well developed at 200 m. Along the Río Cuchujaqui, the forest is well developed at 250 to 400 m. Just to the south in northern Sinaloa, the lower limit of TDF is typically this low.

The easiest field character for distinguishing TDF from thornscrub is the canopy height relative to the columnar cacti. The cacti are beneath the canopy in the forest but emergent in thornscrub. TDF also reaches a greater height with larger trees dominant, achieves continuous coverage, has more mesic species with larger leaves, and contains fewer but larger and taller, thorny and succulent plants. Many species have leaves or leaflets of comparatively large size without the dense hairs and epidermal thickening of leaves in arid environments; however, these leaves are only present in the warm, moist, summer monsoon season.

TDF in the Alamos area has a complicated structure with five different layers or strata: two in the canopy and three in the understory. The tallest trees at 15 m or more above the ground are emergent, widely spaced individuals or stands that represent the northern expression of the *tall tree forests* of southern Mexico and Central America. *Conzattia multiflora* (palo joso de la sierra), an elegant forest giant rising 20 m or more with white bark, is the only common emergent tree (fig. 3.4). Farther south in Mexico, where it is called *guayacán*, a name restricted to *Guaiacum coulteri* in Sonora, this tree is one of the few important timber trees in TDF (Bye 1995). The upper canopy is mostly well developed in forests restricted to slopes of 40° to 60° above 600 m in the Sierra de Alamos but is absent from forests in lower areas, including the Río Cuchujaqui.

The remainder of the nearly continuous canopy is composed of numerous codominant trees that average about 12 m in height (fig. 3.4). In Krizman's (1972) study area at 524 m in the Sierra de Alamos, a relatively mesic site chosen partly because of the presence of large individual *C. multiflora* and *Pachycereus pecten-aboriginum* (etcho), there were 12 species of trees on a 0.1-ha plot and four others nearby. With the exception of *Lysiloma divaricatum*, a legume that has a relatively slender trunk, bark that peels into vertical rectangular plates, and a crown shaped like an inverted cone, most of the trees had obovate crowns. Of 20 species of trees in the area, 90% had broad leaves. Only *Bursera grandifolia* (palo mulato), *Hintonia latiflora* (copalquín), and *Tabebuia impetiginosa* (amapa morada) had the elongate "drip tip" char-

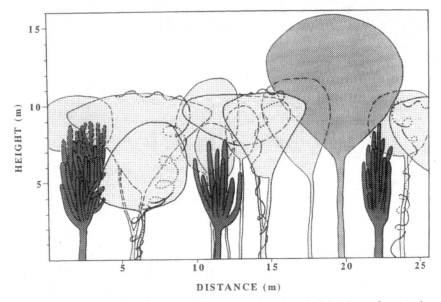

FIGURE 3.4. Cross section of upper and lower canopies of tropical deciduous forest in the Sierra de Alamos, Sonora (modified from Krizman 1972). The upper canopy is represented by a single *Conzattia multiflora* (medium stipple). Trees in the lower canopy (light stipple) include *Bursera stenophylla*, *Ceiba acuminata*, *Ipomoea arborescens*, *Lonchocarpus hermannii*, and *Lysiloma divaricatum* (conical crowns) and *Tabebuia impetiginosa* (circular crown). Columnar cacti (dark stipple) are *Pachycereus pecten-aboriginum* and *Stenocereus thurberi*. Lianas are *Heteropterys palmeri*, *Ipomoea bracteata*, and *Rhynchosia precatoria*.

acteristic of plants in wetter tropical forests. Most (75%) of the trees had relatively smooth, thin bark with or without paperlike exfoliation. Compared to those in oak woodland or conifer forests, the percentages of trees with soft wood are much higher in TDF. The wood ranges from hard and dense, as in *Chloroleucon mangense* (palo fierro), *Haematoxylum brasiletto* (brasil), and *T. impetiginosa*, to moderately soft in *Bursera* spp., to very soft in *Ipomoea arborescens* (morning glory tree, palo santo).

Important tree cacti in the canopy include *Pachycereus pecten-aboriginum*, *Pilosocereus alensis* (old man cactus, pitahaya barbona), *Stenocereus montanus* (sahuira), and *S. thurberi* (pitahaya, organpipe cactus). Lianas such as *Gouania rosei* (huirote de violín), *Heteropterys palmeri*, *Ipomoea bracteata* (jícama), *Marsdenia edulis* (mábim), and *Rhynchosia precatoria* (rosary bean, chanate pusi) commonly reach the canopy.

Krizman recognized three strata in the understory: tall shrubs and small trees up to 5 m, shrubs up to about 1 m (fig. 3.5), and herbs near the ground

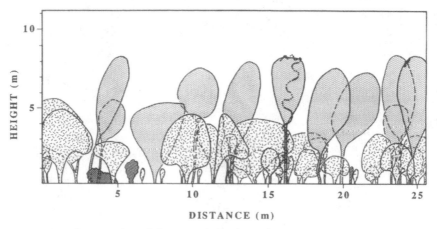

FIGURE 3.5. Cross section of the tropical deciduous forest understory in the Sierra de Alamos, Sonora (modified from Krizman 1972). Short trees (medium stipple) include *Brongniartia alamosana, Ceiba acuminata* (young), *Hintonia latiflora, Senna atomaria*, and *Willardia mexicana* (young). The vine is *Nissolia microptera*. Shrubs (dashed) include *B. alamosana* (young), *Croton fantzianus, Erythroxylon mexicanum, Jatropha malacophylla, Lonchocarpus hermannii* (young), *Montanoa rosei, Randia echinocarpa, S. atomaria* (young), and *S. pallida*. Cacti (dark stipple) are *Opuntia thurberi* and *O. wilcoxii*.

that are mostly present in the summer rainy season. Important shrubs in the shrub-tree layer were *Croton fantzianus* (vara blanca), *Jatropha malacophylla*, and *Montanoa rosei*. The dominant in the lower shrub layer was *Ambrosia cordifolia* (Sonoran bursage); *Senna pallida* (flor de iguana) was common.

In Krizman's plot, the combined total of 238% coverage reflects the overlap of layers with each other. Within the strata, the coverage, a measurement of the amount of ground covered by plants along 220 m of transect lines, ranged from nearly continuous in the lower canopy to about 50% in the shrub-tree layer and 65% in the shrub layer. The vegetation density had the opposite pattern, with increasing numbers of smaller plants per 0.1 ha at lower levels: i.e., emergent stratum = 1, canopy = 104, shrub-tree layer = 239, and shrub layer = 1,573. The total density was 191,700 perennial plants per hectare. This is somewhat denser than Gentry's (1982) estimate of 75,000 to 123,000 plants per hectare. The most common plant, the numerical community dominant, was not a structurally dominant tree but the shrub *Ambrosia cordifolia*, with 1,384 individuals. This is similar to desertscrub communities, where *A. deltoidea* and *A. dumosa* (white bursage) are dominants in the Arizona Upland and Lower Colorado River Valley, respectively.

Although Krizman's dissertation is an excellent study of a TDF stand near

Alamos, the community composition and structure of this forest vary greatly depending on slope, aspect, elevation, substrate, and topographic position. In lower, drier areas, the diversity of trees is somewhat less, and *Lysiloma divaricatum* forms most of the canopy. In the dry season, it is leafless, and light easily passes through its many twiggy branches. For most life-forms, the decrease in species diversity in lower elevation forests on gentler slopes is quickly reversed in the rich canyon habitats of the Río Cuchujaqui.

In a longevity study at Chamela, Stephen P. Bullock (pers. comm. 1994) found that it was difficult to determine the age of trees in TDF. Most appeared to be 50 to 200 years old; a few hardwoods, such as *Tabebuia impetiginosa*, were perhaps as much as 300 years old. These ages contrast with those of oaks or conifers, which typically live 500 years or more. Around Alamos, forests appear to recover from traditional slash-and-burn agriculture in 70 to 100 years. Assessments of the maturity and degree of disturbance of today's forest should take into consideration that few of the trees were alive when the Jesuits built the first church 300 years ago, and the forest could have been through several successional cycles.

The Seasons of Alamos

PHENOLOGY

Water is the primary limiting resource in TDF, where the phenological patterns are stronger than in wet forests (Bullock 1995; Holbrook et al. 1995). Phenological responses are both vegetative (leaves, root or stem growth, etc.) and reproductive (flowers and fruits; Bullock 1995). Here, as in thornscrub and desertscrub, most plants "avoid" drought in one way or another. Phenological responses differ among life-forms and within species because of microhabitat variability in soil moisture (Bullock 1995; Medina 1995).

The primary responses to moisture changes are leaf abscission at the beginning of the dry season and fine root growth and leaf bud break after the first major rain (Bullock 1995; Holbrook et al. 1995). The cumulative transpirational demand imposed by leaves far exceeds the storage volume of even a large tree. It is therefore much easier to be a deciduous species because evergreen trees have thicker and heavier leaves, deeper roots, and a greater investment in xylem transport systems (Holbrook et al. 1995). Deciduous trees respond to rainfall much faster than do evergreens.

The phenological patterns are somewhat more complex in the southern Sonoran forests because of the secondary "winter" rainfall from Pacific fronts. Although the primary wet season of about 3 months beginning in July can be extended by fall tropical storms and winter rains until January or February, March through early July is typically dry. Many deciduous plants drop their leaves promptly at the end of the primary rainy season in late September, whereas others retain part of their leaves until the beginning of the dry season in January or February.

The flowering of deciduous trees in the dry season is essentially uncoupled from available soil moisture. Full hydration of the terminal branches is required for trees to flower. Dry season reproduction is primarily achieved through the use of water stored in the trunk. In some cases the stem water content of trees is higher than that of associated arborescent cacti (Holbrook et al. 1995)! In general, stem diameters swell and shrink by tracking rainfall. Most trees flower less than 6 weeks (Bullock 1995).

At the beginning of the rainy season, deciduous trees respond first by rapid (within 3 days) growth of fine roots, which help with nutrient uptake, followed by leaf buds breaking within 1 week (Bullock 1995; Cuevas 1995; Holbrook et al. 1995). The leaf buds are not really dormant but in a "state of drought-imposed rest" (Holbrook et al. 1995). Productivity of the roots can match the aboveground production (Martínez-Y. 1995). A few wet-season deciduous species, such as *Jacquinia macrocarpa* (sanjuanico), have sclerophyllous leaves in the dry season.

COLORS OF THE FOREST

The most dramatic change in the color landscape occurs at the beginning of the summer monsoon in July, when the forest changes from *monte mojino*, the distinctive, reddish gray-brown color of the dry season, to *monte verde*, the bright green of new leaves. Gentry (1982) poetically described the "long spring dry season forest" as a "dreary scene: a naked infinite host of trunks and branches, spreading interminably over the volcanic hills and mesas, bared to the fiery sun, under which the last leaf seems to have withered and died. He who walks this land in the month of May walks with a parched throat." The *mojino* color is predominantly due to the reddish brown bark and persistent pods of *Lysiloma divaricatum* and *Brongniartia alamosana* (palo piojo) and the leaves of *Croton fantzianus* and *C. flavescens* (vara prieta).

In July the relative calm of the TDF is shattered by monsoonal thunderstorms with strong winds and heavy, intense rain. Overnight the forest becomes browner as the winds of the first storm strip the last yellow leaves clinging to the trees. Within 7 to 10 days, leaf buds erupt, metamorphosing skeletal trees into a dense verdant forest. The bright green of new leaves darkens as they mature over the next few weeks. Even then, the green of the forest is far from uniform, with especially dark green leaves on *Agonandra racemosa* (palo verde, matachamaco), *Ficus* spp., *Guaiacum coulteri* (guayacán), and *Jacquinia macrocarpa*. By this time, the fruits of trees with showy dry season flowers have ripened, allowing the seeds to germinate in the warm, moist conditions that enhance the chances of survival of the seedlings. With new leaves, the trees are again able to produce food through photosynthesis and channel energy into growth.

Haematoxylum brasiletto, a distinctive, small, leguminous tree with a strongly fluted trunk, blotched gray and brown bark, and reddish heart wood, is one of the few plants with showy flowers in the summer rainy season. Although some individuals begin flowering as early as March, most plants have bright yellow flowers in August. The pink and white flowers of legumes, especially *Mimosa palmeri* (chopo), are conspicuous against the forest green. *Hintonia latiflora* (copalquín) is a locally common tree with renowned medicinal properties (Yetman et al., this volume) and with conspicuous, elongate, pale green, club-shaped, ribbed buds and large, delicate, trumpet-shaped white flowers in August and September.

By September, herbs and grasses choke the understory, and vines cover shrubs and trees. A knee-high layer of herbs covers the ground in the forest understory. Flowers of the various morning glory vines, or trompillos, range from pink, purple, red, and orange in *Ipomoea* to scarlet in *Operculina pteripes* and white in *Jacquemontia* and *Merremia*. At Arroyo El Mentidero and many areas near Alamos, *I. pedicellaris* forms spectacular curtains of large pink-purple flowers. *Schizocarpum palmeri* (bitachera) is a common cucurbit vine with bright yellow, star-shaped flowers and triangular fruits.

As the summer monsoons taper off in mid- or late September, the forest begins its secondary dry season with some typical "fall" colors. The leaves of various *Bursera, Erythrina flabelliformis* (coral bean, chilicote), *Fouquieria macdougalii* (tree ocotillo, jaboncillo), *Ipomoea arborescens, Jatropha cordata* (torote papelío), and *J. malacophylla* turn yellow and begin to fall. *Bursera grandifolia* and *Caesalpinia platyloba* (palo colorado) turn maple red, and *Cochlospermum vitifolium* (buttercup tree, palo barril) turn brilliant red and yellow. For these trees, the deciduous season begins at this time and not at the end of the winter rainy season, whereas the leaves of *Tabebuia impetiginosa* and most of the legumes remain green. Through the fall and winter, the partially deciduated forest is a mosaic of greens and browns. The final shift to the dry-season monte mojino comes at the end of the winter rains in January or February.

In the dry season from January to July, the appearance of the forest changes dramatically every 6 to 8 weeks, the Sonoran Desert "winter-spring" and tropical dry seasons combined. Showy flowers, so conspicuous against the monte mojino, rivet the attention of visitors to particular tree species. Most of them disappear into the obscurity of the forest after their turn on stage.

From October or December to April, *Ipomoea arborescens*, "a spectacular tree 7–10 m high, smooth pale bark like the hide of a hippopotamus" holds "a high, thin spread of white corollas like stars against the morning sky. These stars soon fall upon the ground, where deer eat them" (Gentry 1942). The small size of local Sinaloan whitetail deer *(Odocoileus virginianus* subsp. *sinaloae)* allows them to move in dense TDF with ease.

Beginning in December and lasting through February or April, brilliant torches of magenta *Tabebuia impetiginosa* flowers illuminate the landscape. From February to April, peaking in March, *Cochlospermum vitifolium* is especially showy with its crown of huge, waxy, bright yellow flowers held against the sky atop large trees with smooth, silvery gray trunks. In March *Pseudobombax palmeri* (cuajilote) is conspicuous on cliffs above El Guayabo with its elegant white flowers with many erect stamens and pendulous straplike petals against the smooth, ruby-colored bark (fig. 3.1). March is also the time when oak woodlands of the Sierra de Alamos are most easily discerned from a distance because the leaves turn reddish orange and brown before they fall and the trees flower.

Cordia sonorae (palo de asta) is a slender tree with gray bark that is common and widespread, typically growing in groves in the Alamos area but less so along the Río Cuchujaqui. In early April, it has clusters of large white flowers that soon turn the color of coffee. In mid- to late April, *Senna atomaria* (palo zorrillo), a common small tree with abundant yellow pea flowers, brightens the landscape.

In May the dark green leaflets of *Guaiacum coulteri* are covered by sprays of intense purple-blue flowers. In late May *Lonchocarpus hermannii* (nesco) is clothed with masses of magenta flowers. The rest of the year this common tree with its smooth gray bark and well-muscled trunk blends into the forest.

By early June *Ceiba acuminata* (kapok, pochote) is in full bloom with large straplike white flowers. This tall tree with a broad canopy is readily distinguished by the heavy, laminated, pyramidal thorns on its trunk. The fruits develop through the fall and persist after the leaves fall, bursting into pendulous white cottony balls of fluff at least until March. The fibers have been used to make pillows, cushions, and life preservers.

In the heat of mid- to late June, *Brongniartia alamosana*, a common small legume tree that typically grows in groves, displays a peculiar, dark, brownish maroon flower. *Plumeria rubra* (frangipani, cascalosúchil), an especially common shrub on the ledges and shelves above Arroyo El Mentidero, has very large white flowers that appear on leafless, candlelike stems in June and July. Unlike cultivated forms of this species from the more tropical portions of its extensive range, the flowers are not strongly fragrant.

A number of trees and shrubs flower when the forest is most open in the hottest, driest, most desertlike conditions of May and June. They typically have small flowers, likely pollinated by bees and flies, and belong to such families as Erythroxylaceae, Rhamnaceae, Sterculiaceae, and Ulmaceae. This group includes *Celtis iguanea*, *C. pallida* (desert hackberry, cumbro), *C. reticulata* (netleaf hackberry, cumbro), *Colubrina triflora*, *Erythroxylon mexicanum* (momoa), *Guazuma ulmifolia* (guásima), and *Karwinskia humboldtiana* (cacachila).

In June and July, *Lysiloma watsonii* (tepeguaje) is visible in unsuspected numbers, not because of its flowers but because of its conspicuous, bright green new foliage. It leafs out well in advance of the summer rains and ahead of its more common relative, *L. divaricatum*. When trees break leaf buds before the rains begin, a cue such as photoperiod may be involved (Holbrook et al. 1995).

The swollen-trunk torotes *(Bursera* and *Jatropha)* have tiny flowers that are often difficult to see and certainly not showy. A patient observer would find that the first summer rains in July trigger flowers in *B. fagaroides* (torote de venado), *B. lancifolia* (torote copal), *B. laxiflora* (torote prieto), *B. penicillata* (torote acensio), *B. stenophylla* (torote copal), *Fouquieria macdougalii*, and *Jatropha cordata* as their new leaves simultaneously turn the hills a jade green.

MICROCLIMATES OF THE FOREST

Environmental conditions related to the wet and dry seasons are extremely important in TDF microclimates, overriding many climatic changes related to the time of year (Krizman 1972). Throughout the year, nighttime temperatures are warmest in the tree crowns. In contrast, there are strong seasonal differences in the daytime temperatures. The hottest temperatures are in the leafy canopy in summer but at ground level in the dry spring, reflecting profound seasonal differences in the understory microclimates. On June 20, 1966, daily temperatures ranged from 24° to 65°c near the ground in Krizman's Sierra de Alamos study area. On July 19, after the rains had begun and the forest leafed out, the temperature range at the same site was 24° to 32°c—a drop of 33°c in the maximum daytime temperature and a 73% reduction in the daily temperature fluctuation!

The moisture regime is also profoundly influenced by the seasons (Krizman 1972). Mean daily humidity is lowest (31%) in the spring dry season, when soil moisture is far below the permanent wilting point of plants. In the summer monsoon season, humidity increases to 86%, soil moisture is constant, and transpiration is greatly reduced.

The amount of light passing through the canopy is the key factor influencing understory microclimates in the forest. In the dry season, the leafless trees provide little shade; about 90% of the incoming light reaches the ground directly. The soil and air become hot and dry, desiccating the understory herbs that flourished in cool December and January rains. With the exception of the few evergreen species, cacti, and torotes with greenish bark, photosynthesis ceases for the duration of the dry season. Dry season flowers and fruit on leafless trees and terminal stem growth are produced on stored energy reserves (Bullock 1995; Holbrook et al. 1995).

In contrast, the dense foliage of the trees, shrubs, and herbs in the canopy and lower vegetation strata intercept more than 95% of the sunlight in late summer. Heat is distributed in the forest through indirect radiation and convection. The warm, moist, and remarkably constant microclimates of the shady understory encourage rapid and vigorous plant growth.

The yellow and red colors in the forest in late September indicate not only the beginning of the fall season but also the transition to fall forest microclimates. As the torotes and other trees shed their leaves, more light passes through the canopy to heat the soil and dry the air. These drier, more open microclimates persist until the legumes finally drop their leaves and the dry season begins in January. On December 22, Krizman recorded a surface temperature of 12°c; lower temperatures are expected during the winter rains.

The seasons in the forest understory thus begin with an extended version of the hot, dry, bleak foresummer of the Sonoran Desert from February to July. These climatic conditions are followed by tropical humidity, shade, warmth, and luxuriant herb growth during the summer rains, and then by a temperate winter-spring with Mohave and Sonoran Desert wildflowers in December and January.

CYCLES OF THE RIVER

Each year during the summer rainy season and less often during fall or winter storms, the middle stretches of the Río Cuchujaqui are subjected to flooding, which tends to rip out much of the vegetation and generally rearranges the topography of the canyon bottom. As a result, much of the canyon has a rocky or gravelly bed or exposed bedrock. The regular scouring prevents the establishment of dense perennial vegetation on the canyon floor. Where bedrock is at or near the surface, permanent pools are formed, even in those stretches of the river without perennial flow in the dry season.

The floods enrich the canyon habitat as seeds, aquatic animals and plants, organic materials, and nutrients are transported from upland portions of the watershed. The extensive gravel bars left by the annual floods provide habitat during the dry season for numerous species of sun- and moisture-loving annual and perennial herbs, many of which could not maintain themselves along the river if the floods stopped and more trees colonized the canyon bottom. Some of these are typical disturbance plants, whereas others appear to be waifs barely maintaining small populations. Some are species that can tolerate light shade; these thrive on the gravel bars every year but grow under the leafless trees on dry slopes only after unusually wet winters.

Like the forest, the river habitats change dramatically in the dry season. Open stretches dry up; permanent water and aquatic fauna are restricted to bedrock pools or shady *Taxodium distichum*-lined channels. The deeper pools with strong thermoclines serve as seasonal refuges for larger fish, even though there is little circulation; algae build up in warm surface water overlying cooler bottom water. In exposed shallow areas, *Ludwigia peploides* (yellow water weed) locally thrives in water that is too hot for most animals and plants. In areas with more continuous flow, low water exposes rocks covered with the dwarf crustose *Oserya* cf. *coulteriana*.

In several areas, notably at the Las Lajitas crossing, the river forms elongate quiet-water ponds with muddy bottoms overlying the rocky gravel streambed. The 5 or 6 months of the dry season are long enough for these obligate aquatics to develop: *Echinodorus berteroi* (burhead, hierba del manso del agua), *Heteranthera limosa*, *Lemna aequinoctialis* (duckweed), *Marsilea vestita* (water clover), *Najas guadalupensis*, *Potamogeton foliosus*, *P. nodosus* (pondweeds), and *Zannichellia palustris* (common poolmat). Other aquatic plants are not common in the river probably because of the annual floods and the scarcity of quiet pond areas. *Eichhornia crassipes* (water hyacinth) is a floating aquatic with attractive lavender flowers that is native to Central and South America and is common in the Río Mayo and irrigation canals near Navojoa. In 1992 an unsuccessful introduction was made at the Güirocoba crossing.

The aquatic habitats of the river thus change markedly throughout the year, tracking the seasons. In the summer rainy season and occasionally in the winter, the river is flooded with high-energy flow throughout the study area. As the floods recede, the river settles to its normal stream-canyon pool mode. In the spring dry season, the river is further reduced to stretches of shallow stream and dry riverbed between deep pools and quiet ponds.

Flora of the Río Cuchujaqui

GENERAL FLORISTIC ANALYSIS

Our study of the Río Cuchujaqui flora yielded 740 taxa (736 species plus 4 additional subspecies or varieties), 429 genera, and 115 families (appendix 3.1). One species *(Bouteloua eludens* Griffiths; Beetle M-9169) was excluded because the locality data (south of Alamos, Río Cuchujaqui, October 1983) is dubious (John R. Reeder, pers. comm. 1993). Only 46 species (6.2%) are introduced. The Río Cuchujaqui collections have been incorporated into a revision of Río Mayo plants (Martin et al. 1998; Yetman et al. 1995).

A total of 496 taxa (67%) in the Río Cuchujaqui flora were previously reported for the Río Mayo region under the current or other names by Gentry (1942). Although the Alamos area, which is mostly in the Río Fuerte drainage, was only peripherally included in his study area, the addition of 243 taxa to the Río Mayo flora is significant. Gentry made 3,200 collections representing 1,276 species and varieties from various habitats, including TDF north and northeast of Alamos. Of the new additions, only 29 species are exotics that probably entered the area since 1940. Others are in groups, such as grasses (48) and ferns and fern allies (11), that were less thoroughly collected by Gentry. Still others reflect little previous effort on the riverine and pond habitats of the Río Cuchujaqui, notably obligate aquatics (13) and plants of moist streamside soils (19), especially Cyperaceae (8) and Juncaceae (1). Thirty of the new grasses live near or in water. Most of the additions are simply the result of concentrated collection efforts in a local area. Eighteen species in the flora were first reported for Sonora by Van Devender et al. (1995).

A few additions to the Río Mayo flora are new taxa. *Bouteloua quiriegoensis*, an annual grass virtually restricted to the Río Mayo region, was a new species (Beetle and Johnson 1991). The descriptions of *Euphorbia alatocaulis* and *E. dioscoreoides* subsp. *attenuata* included specimens from the study area (Steinmann 1996; Steinmann and Felger 1997). *Cardiospermum cuchujaquense*, only known from Arroyo El Mentidero and the La Ranchería crossing of the Río Cuchujaqui, is closely related to South American species (Ferrucci and Acevedo-Rodríguez 1998). Three species in the genera *Galactia, Haematoxylum*, and *Tragia* are undescribed.

Although the vegetation is TDF structurally dominated by trees and tree-like cacti, the flora is overwhelmingly dominated by herbs (65.4%), including various dicots (49.1%), grasses and sedges (14.1%), vines (5.3%), and aquatics (2.2%). A total of 33.3% of the terrestrial herbs are annuals. Only 31.9% of the species are woody: trees (9.1%), shrubs (13.8%), subshrubs and parasites (5.0%), and woody vines (4.1%). Cacti and other succulents account for only 2.2% of the flora. Only 6.2% of the flora are non-native introductions.

TREES

Trees are the structural dominants in TDF, even though trees and arborescent cacti compose only 9.1% of the flora. Forest trees that are often present on the sierra slopes include *Agonandra racemosa, Brongniartia alamosana, Chloroleucon mangense, Croton fantzianus, Haematoxylum brasiletto, Ipomoea arborescens, Jacquinia macrocarpa, Lonchocarpus hermannii, Schoepfia schreberi* (palo cachora), *Sebastiania pavoniana* (Mexican jumping bean, brincador), *Senna atomaria, Tabebuia impetiginosa*, and *Wimmeria mexicana* (algodoncillo). Torotes with swollen, succulent trunks and light yellow, green, blue-green, or coppery red, papery bark are common. This group includes *Bursera grandifolia, B. laxiflora, Fouquieria macdougalii*, and *Jatropha cordata*. Occasional tall *Ceiba acuminata* stand out in the forest. In the dry season, the dark green forms of arborescent columnar cacti, such as *Pachycereus pecten-aboriginum* and *Stenocereus thurberi*, are easily seen but disappear into the green camouflage of the forest trees in the rainy season. In the El Guayabo area, *Pilosocereus alensis* and *S. montanus* are also present.

Trees that are more frequently found in canyon bottoms include *Albizia sinaloensis, Celtis reticulata, Guazuma ulmifolia, Havardia mexicana* (palo chino), *Leucaena lanceolata* (guaje, palo bofo), *Lysiloma watsonii, Prosopis glandulosa, Sideroxylon tepicense* (tempisque), and *Vitex mollis* (uvalama). A few trees along the river, such as *Leucaena leucocephala* (guaje) and *Psidium guajava* (guava), are introductions from tropical America. *Parkinsonia aculeata* (Mexican palo verde, guacoporo) is so widespread in the New World tropics and so successful in riparian and pseudoriparian habitats (disturbed roadsides, etc.) that it is difficult to know if it is native in the Alamos area.

Other trees, such as *Diospyros sonorae* (Sonoran persimmon, guayparín) and *Pithecellobium dulce* (guamúchil), are closely associated with human settlements. Both are handsome shade trees with edible fruit. Both have Mayo names (caguo rara and macochini), indicating that these people have been familiar with the plants for a long time. The situation for *D. sonorae* is puzzling because the species is known only from Sonora and Sinaloa. How could a regional endemic be introduced? *P. dulce* was long thought to be a native of the East Indies or India. McVaugh (1987), however, reported that it is a widespread New World tropical species that was carried on Spanish galleons to the Philippines and then to India, where it was "discovered" by botanists and described.

Several species that are typically shrubs in the study area were trees at La Ranchería: *Celtis pallida, Phaulothamnus spinescens*, and *Solanum erianthum* (sosa).

SHRUBS

Shrubs compose 13.8% of the flora. Common species in the TDF understory include *Croton flavescens, Jatropha malacophylla, Karwinskia humboldtiana, Randia echinocarpa* (papache), and *Senna pallida. Ambrosia cordifolia* is a common shrub in disturbed forest understory habitats and is an important pioneer species after clear-cutting. Typically, *Acacia cochliacantha* (boatthorn acacia, güinolo), a short-lived shrub or small tree that thrives on disturbance, is the next stage in the succession back to TDF, often in relatively continuous stands. However, when *Pennisetum ciliare* (buffelgrass), a coarse African grass, is planted in the cuts, recurrent fires and nutrient deficits in the soils effectively prevent succession.

Justicia candicans and *Tetramerium abditum* (both rama de venado) are very similar red-flowered, shrubby acanths that are common in TDF in rocky canyons. *Abutilon abutiloides* (malva), *Bastardiastrum cinctum* (malva blanca), *Sida aggregata* (rama lisita), and other shrubby mallows are common in disturbed areas in canyon bottoms. These species grow in a few areas: *Acacia occidentalis* (catclaw, teso), *Bernardia viridis, Caesalpinia pulcherrima* (red bird-of-paradise, tabachín), *Dalea scandens* (escoba), *Hyptis albida* (desert lavender, salvia), *Indigofera suffruticosa* (añil), *Lagascea decipiens, Lantana camara, L. velutina* (all confiturilla), and *Mimosa palmeri. Vallesia glabra* (sitavaro), a common shrub in coastal thornscrub, was found only at the La Ranchería crossing.

Capsicum annuum (chiltepín) is a weak shrub that may climb to more than 2 m by leaning on other shrubs. Its small reddish or orange berry is a very hot, spicy condiment that is widely harvested, sold, and eaten. The locations of plants near dwellings are well known. The Mayo recognize populations with berries of different potency. Cócorit, a Yaqui village on the coastal plain, bears the Cáhitan name for the plant.

SUCCULENTS

Several cacti are important structural elements in the forest, even though perennial succulents compose only 2.3% of the flora. *Stenocereus alamosensis* (sina), a sprawling "galloping cactus" with red tubular flowers in July that is common in thornscrub, was seen only at the La Ranchería crossing. Other succulents in the forest include *Agave vivipera* (maguey), *Mammillaria grahamii* (fishhook cactus), *Opuntia pubescens* (siviri chucha), *O. thurberi* (siviri), and *O. wilcoxii* (prickly pear, nopal). *Ferocactus pottsii* (barrel cactus, biznaga), *Hechtia montana* (mescalito), and *Mammillaria standleyi* (biznaguita) grow mostly on cliffs or steep rocky slopes. *Pereskiopsis porteri* is occasional but widespread in the tropical forest understory along the Río Cuchujaqui. A large patch of *Aloe barbadensis* (aloe vera, sábila), a South African succulent commonly cultivated for its medicinal properties and attractive yellow flowers, was found on a roadside near Arroyo El Mentidero.

VINES

Vines, called *enredaderas, huirotes*, or *trepadoras*, are common in the TDF (69 species, or 9.3% of the flora). They range from woody vines (4.1%), including the stout lianas, to herbaceous perennials (2.7%) and annuals (2.6%). The lianas *Gouania rosei, Ipomoea bracteata*, and *Rhynchosia precatoria* are mostly found in the forest, where they climb to the canopy in search of light (fig. 3.4). In a few places in the shady forest understory, *Macfadyena unguis-cati* (catclaw vine) clings to rocks and trees with its three-pronged claws. Although we seldom saw flowers, these diminutive vines are the ancestral form of the widespread cultivated variety with larger leaves and showy yellow flowers.

Most of the other vines are most common in clearings, along roads, and in open canyons. Vines are especially common in the canyon riparian habitats along the Río Cuchujaqui. Common woody vines include *Callaeum macropterum* (gallinitas, batanene, matanene), *Cocculus diversifolius* (snailseed), *Dioscorea convolvulacea, Serjania mexicana* (huirote de culebra), and milkweeds (*Marsdenia edulis, Matelea petiolaris, Metastelma arizonicum*). *Cynanchum ligulatum* and three species of *Sarcostemma* (Asclepiadaceae) and two species of *Merremia* (Convolvulaceae) are large vines that are not especially woody. *Echinopterys eglandulosa* is a common woody vine with showy spikes of yellow flowers from March to May that climbs 2–3 m in trees and shrubs. Surprisingly, near Guaymas in the Sonoran Desert the same species is a shrub.

Vines restricted to riparian habitats along the river or in Arroyo El Mentidero include *Byttneria aculeata* (almohada de culebra), *Cissus verticillata* (huirote de agua), *Sarcostemma clausum*, and *S. cynanchoides* (milkweed vines). *Diodia sarmentosa* is a tropical vine with small white flowers on herbaceous stems that trail across the ground from a woody base. In a few places along the river, *Cryptostegia grandiflora* (bejuco), a woody vine or scandent shrub with a large white and lavender flower, native to India and widely cultivated as an ornamental, forms dense thickets.

Dalechampia scandens is a widespread herbaceous perennial vine in Euphorbiaceae whose leaves and inflorescences bear a strong resemblance to the uncommon *Passiflora foetida* (passionflower, talayotes). *Aristolochia quercetorum* and *A. watsonii* (pipe vines, hierbas de la india) are found in a few places along the river. *Clematis drummondii* (Virgin's bower, barba de chivato), a widespread vine in riparian areas in the southwestern United States and northern Mexico, was found in only three localities along the Río Cuchujaqui.

Herbaceous vines in Convolvulaceae (17 species, 5 genera) and Cucurbitaceae (10 species, 10 genera) abound along the river in the summer rainy season in late August and September, typically covering trees, shrubs, and fences in a frenzy of growth. The genus *Ipomoea* has 11 species in the area; all but *I. arborescens* are vines. *Sechiopsis triquetra* and *Sicyos sinaloae* are annual vines whose large thin leaves form dark green, leafy curtains up to 8 m on fences, shrubs, and trees in only 2 months!

GRASSES, SEDGES, AND RUSHES

Although grasses are structural dominants only in planted buffelgrass pastures (*praderas*), they are important throughout the study area. With 82 species in 37 genera, grasses are second only to legumes in the diversity of species. *Bouteloua* (grama grass) with eight and *Eragrostis* (lovegrass), *Muhlenbergia* (muhly), and *Paspalum* with seven each are the most species-rich genera.

Fully 82.9% of the grasses are native species. Common introduced species include *Cynodon dactylon* (Bermuda grass), *Dichanthium annulatum* (Kleberg bluestem, fuzzynode), *Echinochloa colonum* (jungle rice), *E. crusgalli* (barnyard grass), *Pennisetum ciliare*, *Rhynchelytrum repens* (natal grass, zacate rosado), and *Sorghum halepense* (Johnson grass).

The grass flora includes species that are widespread in the southwestern United States as well as more tropical areas to the south. *Bothriochloa barbinodis* (cane beardgrass), *Bouteloua chondrosioides* (sprucetop grama), *B. curtipendula* (sideoats grama), *B. radicosa* (purple grama), and *Eragrostis intermedia* (plains lovegrass) are typical desert grassland species in the southwestern United States. *Aristida ternipes* var. *ternipes* (spider grass), *Bouteloua aristidoides* (six-weeks needle grama, zacate saitilla), *B. barbata* (six-weeks grama), *Cathestecum brevifolium* (false grama, zacate borreguero), and *Heteropogon contortus* (tanglehead) are common Sonoran Desert grasses. Here the growth form of *A. ternipes* is a 1.5- to 2.0-m-tall erect grass rather than the rounded 0.5- to 1.0-m-diameter clump grass of the Sonoran Desert. *Bouteloua alamosana* and *B. quiriegoensis* are annuals endemic to southern Sonora that are closely related to widespread desert grassland perennials: *B. radicosa* and relatives, and *B. hirsuta* (hairy grama), respectively.

The grasses range from the 3-m-tall streamside *Arundo donax* (carrizo) to delicate annuals. Annuals account for 48.0% of the species, although some of them, such as *Leptochloa mucronata* (six-weeks sprangletop), can reach 1 m in height. In general, the grasses are more diverse in canyon riparian habitats. *Muhlenbergia elongata* (cliff muhly) is a perennial bunchgrass restricted to steep canyon walls. Many species, including *Arundinella palmeri* (cañuela), *Cenchrus* (sandburs, guachaporitos; 3 species), *Dactyloctenium aegypticum* (crowfoot grass), *Digitaria* (3), *Echinochloa* (2), *Eleusine indica* (goose grass), *Eragrostis* (2), *Panicum virgatum* (switchgrass), *Paspalum* (7), *Polypogon monspeliensis* (rabbitfoot grass), and *Setaria parviflora* (stream bristlegrass), are restricted to moist soil in the streambed or near the river.

In the water and wetter areas, various members of Cyperaceae are more common than grasses. These include *Cyperus* (flatsedge, coquillo; 15) *Eleocharis* (spike rush; 3), *Fimbristylis dichotoma*, and *Lipocarpha micrantha*. Surprisingly, no *Carex* (sedges) were found, and *Juncus tenuis* (slender rush; Juncaceae) was found only at El Guayabo. *Typha domingensis* (cattail, tule) was found only in a marshy area on Rancho El Conejo.

In early Tertiary forests 30 to 50 million years ago, the ancestors of living bamboos were broad-leaved understory herbs (Van Devender 1995). In the modern forest, the annuals *Oplismenus burmannii* (zacate cadillo) and *Panicum trichoides* and the perennial *Lasiacis ruscifolia* (negrito) have broad leaves but belong to the more advanced Panicoideae subfamily.

HERBS

Perennial and annual herbs (excluding grasses and sedges) are important (49.1%) in the flora, especially in the forest understory and canyon bottoms. Members of Acanthaceae, with seven species in seven genera, are especially important perennial herbs. One of these, *Henrya insularis* (rama del toro), with its small, pale yellow flowers, forms dense leafy banks in the spring. Other important understory perennial herbs include *Acalypha subviscida, Dalea elata, Hibiscus biseptus* (rose mallow), *Petiveria alliacea* (zorrillo), *Plumbago scandens* (hierba de alacrán, cresta de gallo), *Priva lappulacea* (cordoncillo), and *Salpianthus purpurascens* (jarilla). *Dorstenia drakeana* (baiburilla, pionía) is an herb in Moraceae with large, deeply notched leaves that rise from subterranean root crowns on rocky, shaded forest floors. Its fruits are unusual structures that resemble figs flattened out into oval disks on erect slender peduncles. *Ambrosia confertiflora* (slimleaf bursage, estafiate) and *Parthenium hysterophorus* are common perennial herbs in disturbed soil.

Common perennial herbs in moist soil near the river or nearby riparian habitats include *Blechum pyramidatum, Chaetymenia peduncularis, Elephantopus spicatus, Hydrolea spinosa* (cilantro), and *Tridax procumbens. Bidens sambucifolia* and the annual *Cosmos sulfureus* (tostones) are conspicuous herbs with showy orange flowers that can reach 2 m in height. *Commelina diffusa* and *C. erecta* (dayflowers, hierba del pollo) are common succulent perennial herbs in shady areas and moist sand along the river. *Ricinus communis* (castor bean, higuerilla) almost defies life-form classification by growing as a perennial herb, shrub, or tree. This conspicuous plant of disturbed areas above the river, originally native to the Old World tropics, was not reported by Gentry (1942) and apparently arrived in the Río Mayo region subsequently.

An epiphyte is a plant that grows on another plant and produces its own food by photosynthesis using water and nutrients harvested from the air. Epiphytes are more common in coastal forests with enhanced dew (Gentry 1995). The only epiphytes in the study area are the bromeliad *Tillandsia recurvata* (ball moss, mescalito de árbol) and the orchid *Oncidium cebolleta* (cuerno de chivita). The definition quickly becomes confused because both species are also found attached to rocks. Another bromeliad, *T. elizabethae*, and the orchid *Encyclia trachycarpa* are herbaceous perennials that mostly live on rock surfaces. The only other orchids in the study area *(Habenaria quinqueseta* and *Spiranthes polyantha)* are terrestrial with their roots in soil.

Other perennial herbs are parasites on other plants, taking all of their water and nutrients from their hosts. *Struthanthus palmeri* is a common widespread mistletoe (toji) that lives on various trees, including *Brongniartia alamosana, Guazuma ulmifolia, Prosopis glandulosa, Salix bonplandiana* (Bonpland willow, sauce), and *Taxodium distichum. Phoradendron quadrangulare* (mistletoe, toji) was only found on a single *S. bonplandiana* at El Guayabo. The only annual parasites are a few dodders, including *Cuscuta americana* (golden dodder) on *Haematoxylum* sp. nov. (brasil chino) and *C. boldinghii* on *Dicliptera resupinata* and *Sida* sp. in Arroyo El Mentidero, and *C. potosina* on *Ayenia filiformis* and *Evolvulus alsinoides* in a recently clear-cut forest above El Guayabo.

Pholisma culiacanum is a root parasite that mostly grows underground. Only the fleshy, funguslike heads that bear tiny pinkish white flowers can be seen above the surface. It is closely related to *P. arenarium* and *P. sonorae* of the Sonoran Desert in northwestern Sonora, southwestern Arizona, and southeastern California. *P. sonorae* was the "sand food" historically eaten by the O'odham who lived in the Gran Desierto. Unlike the desert species, *P. culiacanum* is found in rocky understory habitats in TDF and thornscrub, not in sand dunes. The Mayo call it an *hongo,* a general term that refers to all fungi, including mushrooms. They do not eat hongos of any kind!

Ferns, with 15 species (2.0%) in nine genera, can be locally common perennial herbs, especially on rocky soils and cliffs in the shaded forest understory. The resurrection plant, *Selaginella pallescens* (siempre viva), is widespread and common, whereas two spike mosses *(S. sartorii* and *S. sellowii)* were found only in a few locations on volcanic benches above the river.

ANNUALS

There are considerably fewer species of annuals (exclusive of grasses and sedges) in the Río Cuchujaqui flora (29.6%) than in the Arizona Upland flora of the Tucson Mountains (46.0%; Rondeau et al. 1996). However, annuals can be found, at least in riparian canyon bottoms, throughout the year. In late February and March in years with adequate rain, widespread winter-spring desert annuals, including *Daucus pusillus* (wild carrot), *Descurainia pinnata* (tansy mustard, pamita), *Eucrypta chrysanthemifolia, Galium proliferum* (bedstraw), *Lepidium lasiocarpum* (peppergrass, lentejilla), *Parietaria hespera* (pellitory), *Perityle californica, Spermolepis echinata* (scaleseed), and *Sphaeralcea coulteri* (globe mallow), can be common in the forest mixed with abundant regional species, such as *Drymaria glandulosa, Perityle microglossa,* and *Phacelia gentryi* (Sonoran caterpillar weed), a southern relative of the widespread *P. distans.* Succulent annuals with tropical distributions in Commelinaceae, including *Callisia monandra, Tinantia longipedunculata,* and *Tripogandra palmeri,* are scattered in shady habitats along the river. Introduced annuals regularly found along the river include *Anagalis arvensis* (scarlet pimpernel), *Chenopodium murale, Leonotis nepetaefolia* (lion's ears, cordoncillo de San Francisco), *Melilotus indicus* (sour clover, trebolín), *Sisymbrium irio* (London rocket, pamitón), and *Sonchus oleraceus* (sow thistle). Escaped cultivated plants occasionally encountered include *Citrullus lanatus* (watermelon, sandía), *Coriandrum sativum* (cilantro), *Sesamum indicum* (sesame, ajonjolí), and *Solanum lycopersicum* (tomato). More unusual annuals include *Cannabis sativa* (marijuana, mota) and *Papaver somniferum* (opium poppy, amapola), escapees from illicit regional crops.

In September and early October, the summer annuals reach their peak in abundance and biomass and are generally more visible than at other times of the year. Various annual species of *Amaranthus* (2), *Euphorbia* (12), *Ipomoea* (8), *Martynia annua* (aguaro), Compositae (26), Cucurbitaceae (8), and Leguminosae (22) are present in large numbers.

Phytogeography

REGIONAL FLORAL COMPARISONS

To understand the biogeographical affinities of the flora of the Río Cuchujaqui TDF, the distributions of species were analyzed through comparisons with floras from Jalisco on the Pacific coast of Mexico to the south, the Cape region of Baja California Sur, coastal thornscrub in southern Sonora, the Sierra Madre Occidental to the northeast in Chihuahua, and the Sonoran Desert region of northwestern Mexico and southern Arizona (table 3.1). Lott (1993) reported 1,120 species in 124 families in TDF at Chamela, Jalisco (350 km², 0–500-m elevation). Lenz (1992) reported 1,053 species in 131 families in thornscrub, TDF, and pinyon-oak woodland in the Cape region of Baja

Table 3.1. Percentages of Life-Forms in Río Cuchujaqui Flora and Shared Species with Other Floras from Northwestern Mexico and Southern Arizona

FLORA	AREA (km²)	No. TAXA	No. SS	% RC	% LC	% TR	% SH	% HS	% SU	% GR	% GA	% VW	% HB	% HP	% HA	% AQ	% INT
Río Cuchujaqui	46	740				9.1	13.8	5.0	2.2	14.1	6.9	4.1	65.4	21.7	33.3	2.2	6.2
Other floras																	
Chamela	350	1,120	251	34.0	22.4	10.0	15.5	4.8	1.2	13.1	6.0	4.0	60.2	18.3	22.3	1.6	4.8
Cape region	—	1,053	231	31.3	21.9	8.7	11.7	4.3	2.2	18.2	10.4	3.5	67.5	16.0	27.7	2.2	6.1
Coastal thornscrub	5,750	510	298	40.3	58.4	11.7	17.8	4.4	3.4	12.1	7.0	4.4	49.7	14.8	22.8	1.3	8.1
Basaseachi	65	823	151	20.4	18.3	9.3	13.9	4.6	2.6	15.9	6.6	4.0	62.3	18.5	22.5	2.6	4.0
Gran Desierto	4,578	145	19	2.6	13.1	0	21.1	5.3	0	31.6	26.3	0	73.7	5.3	36.8	0	0
Tucson Mountains	4,000	61c	140	18.9	23.0	1.4	13.6	3.6	0.7	22.1	10.0	2.1	74.3	17.1	35.0	0	12.9
Sycamore Canyon	9	624	131	17.7	21.0	3.1	9.9	4.6	0.8	19.1	9.2	3.1	77.1	24.4	29.8	1.5	4.6

Sources: Chamela, Lott (1993); Cape region, Lenz (1992); Coastal thornscrub, Friedman (1996); Basaseachi, Spellenberg et al. (1996); Gran Desierto, Felger (1980); Tucson Mountains, Rondeau et al. (1996); Sycamore Canyon, Toolin et al. (1979).

Note: Column headings are as follows: ss = shared species with Río Cuchujaqui flora; rc = Río Cuchujaqui flora; lc = local flora; tr = trees; sh = shrubs; hs = subshrubs and woody parasites; su = cacti and rosette succulents; gr = grasses and sedges; vw = woody vines; ga = annual grasses and sedges; hb = total annual and perennial herbs, including grasses and sedges; hp = perennial herbs, including ferns but not sedges and grasses; ha = annuals exclusive of grasses and sedges; aq = aquatic herbs; and int = introduced.

California Sur. A list of the plants in coastal thornscrub west of the Alamos area (5,750 km^2, 0–200 m) in southern Sonora contained 510 species in 88 families (Friedman 1996). Spellenberg et al. (1996) reported 823 species in 117 families for pine-oak forest and TDF in the Parque Nacional Cascada de Basaseachi in the central Sierra Madre Occidental in Chihuahua (65 km^2, ca. 850–2,320 m). Felger (1980) found 145 species in 36 families on sand dunes and the Sierra del Rosario in Lower Colorado River Valley desertscrub in the Gran Desierto of northwestern Sonora (4,578 km^2, 0–60 m). The Tucson Mountains in Arizona Upland Sonoran desertscrub in southern Arizona (4,000 km^2, 650–1,430 m) yielded 610 species in 80 families (Rondeau et al. 1996). Sycamore Canyon, a desert grassland/oak woodland/canyon riparian ecotone in southern Arizona (9 km^2, 1,060–1,230 m), contained 624 floral species in 96 families (Toolin et al. 1979). Some differences are inherent in the floristic comparisons because habitats are unevenly represented in the study areas: for example, Chamela and the Río Cuchujaqui have rivers, the Tucson Mountains lack a stream, the thornscrub has coastal mangroves, and the Gran Desierto has massive sand dunes. Other differences are related to variations in regional climate, physiographic setting, and the size and elevational ranges of the study areas.

It was readily apparent in the analyses that the depauperate flora of the harsh, arid Gran Desierto shares few species (19, 2.6%) with the Río Cuchujaqui or any of the other areas. The greatest numbers of species in the Río Cuchujaqui flora were shared with coastal thornscrub (298), Chamela (251), and the Cape region (231; 40.4%, 34.0%, and 31.3%, respectively, of the total number of species in each of these floras). The close relationship between TDF and thornscrub in southern Sonora is not surprising if one considers their geographic proximity. The weaker ties to the Chamela flora reflect important differences between the two tropical deciduous forests. If the woodland floras of the Sierra de la Laguna could be extracted, the Cape region flora would have a higher percentage of shared species. The numbers of Río Cuchujaqui species in the Basaseachi (151), Tucson Mountains (140), and Sycamore Canyon (131) floras were similar, reflecting floristic ties with both Sonoran desertscrub and desert grassland. Surprisingly, the percentages of these floras (18.3%, 23.0%, and 21.0%, respectively) that coexist in the Río Cuchujaqui were not very different from the Chamela value (22.4%). Except for thornscrub, the percentages of shared species were not very high for any of the floras. Apparently, the floras are reasonably rich in most of the areas, and there is a great deal of species turnover in the various communities.

Surprisingly few of the structural dominants of the Río Cuchujaqui TDF are shared with the other floras. Herbaceous plants (including grasses and sedges) are the most important shared life-forms, with the following percentages of shared species: 49.7–67.5% for Chamela, thornscrub, the Cape region, and Basaseachi; 73.7% for the Gran Desierto; and 74.3–77.1% for Sycamore Canyon and the Tucson Mountains. Annuals (exclusive of grasses and sedges) are the most common single life-form with shared species: 22.3–27.7% in the southern floras and 29.8–35.0% in the northern floras. Grasses are somewhat more important among shared species in Arizona (19.1–22.1%) than among those in thornscrub and Chamela (12.1–13.1%).

Only relatively minor differences in the life-forms of shared species could be attributed to differences in vegetation and more tropical climates: i.e., more Río Cuchujaqui trees and perennial vines were shared with Chamela (10.0%, 5.5%) and thornscrub (11.7%, 6.6%) than with the Tucson Mountains (0.7%, 3.1%) and Sycamore Canyon (3.1%, 3.7%). The low percentage of trees (9.3%) shared with Basaseachi primarily reflects the difference between high-elevation pine-oak forest and lowland tropical forest.

The paucity of Río Cuchujaqui trees shared with the Cape region (8.7%) is surprising because both areas support TDF. Although the two forests may appear similar because of some shared forest dominants *(Lysiloma divaricatum, Pachycereus pecten-aboriginum*, and *Stenocereus thurberi)*, the differences in composition are dramatic. The Cape region forests lack many plants that are common in the Río Cuchujaqui forest, including trees (*Acacia occidentalis, Agonandra racemosa, Alvaradoa amorphoides, Brongniartia alamosana, Bursera* [5 species], *Caesalpinia platyloba, Ceiba acuminata, Ficus* [4 species], *Guazuma ulmifolia, Heliocarpus attenuatus, Hintonia latiflora, Lonchocarpus hermannii, Lysiloma watsonii, Sideroxylon persimile, S. tepicense, Tabebuia impetiginosa, Vitex mollis, Wimmeria mexicana*, etc.); columnar cacti (*Pilosocereus alensis* and *Stenocereus montanus)*; and large shrubs (*Celtis iguanea, Erythroxylon mexicanum, Esenbeckia hartmani, Guaiacum coulteri, Jacquinia macrocarpa, Mimosa palmeri, Pisonia capitata, Randia echinocarpa*, etc.).

Considering the distributions of individual species provides yet another way to look at the floristic affinities of the Río Cuchujaqui flora. None of the species occurs in all of the sites. *Heteropogon contortus* (tanglehead) is a widespread tropical grass that was in all of the floras except coastal thornscrub. *Hyptis albida* lives in all areas except Sycamore Canyon. *Datura discolor* (desert thorn apple) was absent only from Basaseachi and Sycamore Canyon. These three are the only Río Cuchujaqui species that were found in both Chamela and the Gran Desierto. A total of 44 Río Cuchujaqui species occurred in both Chamela and the Tucson Mountains. A total of 65 Río Cuchujaqui species were found in both the Cape region and Basaseachi; most of them (72.7% and 72.3%, respectively) are widespread herbs.

In summary, the comparisons of the Río Cuchujaqui flora with seven other local floras in Jalisco and the Sonoran Desert region of northwestern Mexico and southern Arizona yielded some surprising results. Except for the nearby coastal thornscrub and the Gran Desierto, about a fifth of the species of these floras also grew along the Río Cuchujaqui. Most of the shared species were herbs, the dominant life-forms in all of the floras, and not community structural dominants. In the more tropical floras (Río Cuchujaqui, thornscrub, and Chamela), trees and woody vines are more important than in desertscrub or desert grassland. The modest floristic overlap between the Río Cuchujaqui and Chamela (34.0%) indicates that the structural similarities in the TDF in these areas are due to similar dominant life-forms, primarily trees, but the species composition is different.

The closest relationships of the Río Cuchujaqui flora are, not surprisingly, with the adjacent coastal thornscrub. The least affinities are with the sparse Sonoran desertscrub in the Gran Desierto. The floristic affinities with the desert grassland/oak woodland/canyon riparian ecotone in Sycamore Canyon are intermediate and very similar to those with Chamela. A group of tropical species, including *Capsicum annuum, Erythrina flabelliformis, Henrya insularis, Lagascea decipiens, Manihot angustiloba, M. davisiae, Marina diffusa, Rhynchosia precatoria, Tillandsia recurvata*, and many others, reach their northern limits at an elevation of 1,200–1,650 m in the desert grassland/oak woodland transitions of southern Arizona. Their distributions are limited there by freezing temperatures in the woodlands and forests at higher elevations and by heat, aridity, and low humidity in the deserts below (Van Devender et al. 1994).

COMPARISONS WITH OTHER FLORAS

The families with the greatest numbers of taxa in the Río Cuchujaqui flora (*n* = 739) are Leguminosae (11.6%), Gramineae (11.1%), and Compositae (10.1%). Other important families with 40 to 15 species include Euphorbiaceae (5.4%), Malvaceae (3.8%), Solanaceae (3.5%), Convolvulaceae (2.8%), Cyperaceae (2.7%), Rubiaceae (2.2%), and Acanthaceae (2.0%). Species in these 10 families account for 55.4% of the flora. The numbers of species in the largest genera in the flora are 18 (*Euphorbia*), 15 (*Cyperus*), 11 (*Ipomoea* and *Solanum*), 8 (*Bouteloua*), and 7 (*Acalypha, Anoda, Eragrostis, Muhlenbergia, Paspalum*, and *Sida*).

Gentry's (1995) floristic analyses of 25 Neotropical dry-forest floras help to evaluate the Río Cuchujaqui flora. All of the important Río Cuchujaqui families are found on the "top 10" lists for other dry tropical floras. Leguminosae is consistently the most common family, followed by Euphorbiaceae and Gramineae. Compositae is especially important in both the Chamela (Lott 1993) and Río Cuchujaqui floras. The Chamela flora is relatively richer in Euphorbiaceae, Acanthaceae, Convolvulaceae, and especially Bromeliaceae; the Río Cuchujaqui flora, however, is richer in Gramineae and Cyperaceae. The floristic composition of the coastal thornscrub (Friedman 1996) is very similar to that of the Río Cuchujaqui in general, except that Gramineae (9.4%) was the most common family and Cactaceae (4.3%), Nyctaginaceae (2.5%), and Asclepiadaceae (2.5%) were relatively more common. In contrast, the most numerous families in the Cape region, Basaseachi, and the Tucson Mountains were Compositae (10.4%, 18.0%, 17.0%), Gramineae (10.4%, 10.1%, 15.0%), and Leguminosae (8.9%, 7.9%, 8.0%). In the Cape region, Euphorbiaceae was also common (4.7%). Orchidaceae (3.2%), Cyperaceae, Labiatae, Scrophulariaceae (all 2.4%), and Euphorbiaceae (2.3%) were common at Basaseachi.

Gentry (1995) found that the families with the greatest numbers of tree species in 25 dry forest floras were Leguminosae, Bignoniaceae, Rubiaceae, Sapindaceae, Euphorbiaceae, Flacourtiaceae, and Capparaceae. This is in marked contrast to the Río Cuchujaqui flora, in which the families with the most tree species are Leguminosae (2.3%), Burseraceae (0.8%), Cactaceae (0.7%), and Moraceae (0.7%). The Río Cuchujaqui flora also differs from that of Chamela: the most common trees in the latter are in Leguminosae, Euphorbiaceae, Bignoniaceae, and Rubiaceae. There are no Río Cuchujaqui trees in several families that are important in TDF elsewhere: Anacardiaceae, Annonaceae, Apocynaceae, Capparaceae, Combretaceae, Flacourtiaceae, Hippocrateaceae, Nyctaginaceae, and Polygonaceae. The importance of trees in Burseraceae, Cactaceae, Convolvulaceae (*Ipomoea*), and Moraceae and the presence of trees in Fouquieriaceae and Zygophyllaceae in the Río Cuchujaqui flora reflect the relatively dry climates of TDF in southern Sonora.

ACKNOWLEDGMENTS

Victor W. Steinmann, Kristen J. Johnson, and Jessie C. Piper helped in many ways with the identification and processing of plant collections. We thank the many colleagues and associates who helped in the field. Mayo and Spanish common names were provided by David A. Yetman, Ana L. Reina G., and their consultants. We thank the numerous specialists who gave freely of their time and knowledge to help make this flora as accurate and the nomenclature as current as possible. Anne E. Gondor drafted the figures. Frank W. Reichenbacher allowed use of the TDF distribution in his map of biotic communities in North America (Reichenbacher et al, 1998) for figure 3.3. Plants were collected under permits to Thomas R. Van Devender and Stephanie A. Meyer from the Navojoa office of the Secretaría de Agricultura y Recursos Hidráulicos, and to Thomas Van Devender and Alejandro E. Castellanos V. (Departamento de Investigaciones Científicas y Tecnológicas de la Universidad de Sonora) from the Secretaría de Medio Ambiente Recursos Naturales y Pesca in México, D.F.

APPENDIX 3.1

Annotated Checklist of Río Cuchujaqui Plants

---

* = species without voucher specimens.

Local = 1–3 localities in study area;

common = 4–6 localities;

widespread = 7–9 localities.

M = Mayo common names; other common names are English or Spanish.

FERNS AND ALLIES

ASPLENIACEAE

Asplenium pumilum Sw. Local.

MARSILEACEAE

Marsilea vestita Hook. & Grev. Water clover. Local aquatic.

PTERIDACEAE

Adiantum tricholepis Fee. Maidenhair fern, pajarina, cañaguala. Common.

Astrolepis sinuata (Sw.) Benham & Windham subsp. *sinuata* [*Notholaena sinuata* (Sw.) Kaulf.]. Wavyleaf cloak fern. Common.

Cheilanthes bonariensis (Willd.) Proctor [*Notholaena aurea* (Poir.) Desv.]. Local.

Cheilanthes brachypus (Kunze) Kunze. Local.

Cheilanthes lozanii (Maxon) R. Tryon & A. Tryon var. *seemannii* (Hook.) Mickel & Beitel. Local.

Cheilanthes skinneri (Hook.) R. Tryon & A. Tryon. Local.

Notholaena candida (Mart. & Gal.) Hook. var. *candida*. White cloak fern. Local.

Notholaena grayi Davenp. Cloak fern. Local.

Notholaena lemmonii D.C. Eaton var. *lemmonii*. Cloak fern. Local.

Pellaea ovata (Desv.) Weatherby. Local.

SCHIZAEACEAE

Anemia affinis Baker. Local.

Anemia tomentosa (Savighy) Sw. var. *mexicana* (C. Presl) Mickel. Local.

Selaginella pallescens (C. Presl) Spring var. *pallescens*. Resurrection plant, siempre viva, teta segua (M). Common.

Selaginella sartorii Hieron. Spike moss. Local.

Selaginella sellowii Hieron. Spike moss. Local.

THELYPTERIDACEAE

Thelypteris puberula (Baker) Morton var. *sonoriensis* A.R. Smith. Cola de toro. Local.

Gymnosperms

CUPRESSACEAE

Taxodium distichum (L.) J.M.C. Rich. var. *mexicanum* Gordon (*T. mucronatum* Ten.). Mexican bald cypress, sabino, ahuehuete. Widespread tree.

Dicots

ACANTHACEAE

Aphanosperma sinaloensis (Leo. & Gentry) T.F. Daniel. Local subshrub. [H. S. Gentry 854].

Blechum pyramidatum (Lam.) Urb. Common perennial herb.

Carlowrightia arizonica A. Gray. Palo blanco, ánima ogua (M). Widespread subshrub.

Carlowrightia pectinata Brandegee. Local subshrub.

Dicliptera resupinata (Vahl) Juss. Rama del toro, alfalfilla. Widespread perennial herb.

Dyschoriste hirsutissima (Nees) Kuntze. Valeriana del monte, ciática. Local perennial herb.

Elytraria imbricata (Vahl) Pers. Cordoncillo. Widespread perennial herb.

Henrya insularis Nees. Rama del toro, alfalfilla. Widespread subshrub.

Holographis pallida Leo. & Gentry. Ramito chino. Local shrub.

Justicia candicans (Nees) L.D. Benson [*Jacobinia ovata* A. Gray]. Rama venado, palo de venado, muicle cimarrón, ma'aso o'ota (M). Common shrub.

Justicia caudata A. Gray. Local perennial herb.

Ruellia intermedia Leonard. Rama del toro, papachili (M). Local perennial herb.

Tetramerium abditum (Brandegee) T. Daniel. Common shrub.

Tetramerium nervosum Nees. Widespread perennial herb.

Tetramerium tenuissimum Rose. Rama del toro. Common subshrub.

AIZOACEAE

Trianthema portulacastrum L. Verdolaga blanca. Local annual.

AMARANTHACEAE

Alternanthera stellata (S. Watson) Uline & Bray. Widespread perennial herb.

Amaranthus brandegei Standl. Pigweed, bledo, quelite, hué (M). Local annual.

Amaranthus palmeri S. Watson. Pigweed, bledo, quelite, hué (M). Widespread annual.

Gomphrena sonorae Torr. Globe amaranth. Widespread annual.

Iresine calea (Ibáñez) Standl. Zerote de cochi. Local shrub.

Iresine celosia L. Common perennial herb.

Annotated Checklist of Río Cuchujaqui Plants (continued)

Iresine hartmani Uline. Local shrub.

Iresine interrupta Benth. Hierba de jiote. Local shrub.

Tidestromia lanuginosa (Nutt.) Standl. Mochis (M). Local annual.

ANACARDIACEAE

Rhus radicans L. Poison ivy, hiedra. Common woody vine.

APOCYNACEAE

Plumeria rubra L. f. Frangipani, cascalosúchil, candelilla. Common shrub.

Stemmadenia tomentosa Greenm. var. *palmeri* (Rose & Standl.) Woodson. Huevos del toro, berraco (M). Local shrub.

Vallesia glabra (Cav.) Link. Sitavaro (M). Local shrub.

ARISTOLOCHIACEAE

Aristolochia quercetorum Standl. Pipe vine, hierba del indio. Local perennial herb.

Aristolochia watsonii Woot. & Standl. Pipe vine, hierba del indio, quasana jibuari (M). Local perennial herb.

ASCLEPIADACEAE

Asclepias leptopus I.M. Johnst. Milkweed. Local perennial herb.

Cryptostegia grandiflora (Roxb.) R. Br. Bejuco. Escaped ornamental native to India. Widespread woody vine.

Cynanchum ligulatum (Benth.) Woodson. Milkweed vine. Local perennial vine.

Gonolobus cf. *uniflorus* H.B.K. Milkweed vine. Local woody vine.

Marsdenia edulis S. Watson. Milkweed vine, mabe, tonchi (M). Local woody vine.

Matelea petiolaris (A. Gray) Woodson. Milkweed vine. Local woody vine.

Metastelma arizonicum A. Gray (*Cynanchum arizonicum* A. Gray) Shinners. Milkweed vine. Local woody vine.

Sarcostemma clausum (Jacq.) Roem. & Schult. Climbing milkweed. Widespread perennial vine.

Sarcostemma cynanchoides Decne. subsp. *hartwegii* (Vail) Holm. Climbing milkweed, huichori (M). Local perennial vine.

Sarcostemma pannosum Decne. Climbing milkweed. Local perennial vine.

BEGONIACEAE

Begonia palmeri S. Watson. Local perennial herb.

BIGNONIACEAE

Macfadyena unguis-cati (L.) A. Gentry. Catclaw vine, huirote sarnoso, huiroa que cara (M). Common woody vine.

Tabebuia impetiginosa (Mart.) Standl. [*T. palmeri* Rose]. Amapa morada, to'obo (M). Common tree.

BIXACEAE (COCHLOSPERMACEAE)

Amoreuxia gonzalezii Sprague & Riley. Saya mome (M). Local perennial herb.

Cochlospermum vitifolium (Willd.) Spreng. Buttercup tree, palo barril, palo barriga, ciánori (M). Local tree.

BOMBACACEAE

Ceiba acuminata (S. Watson) Rose. Kapok, pochote, baogua (M). Widespread tree.

Pseudobombax palmeri (S. Watson) Dugand. Cuajilote. Local tree.

BORAGINACEAE

Cordia sonorae Rose. Palo de asta, pómajo (M). Common tree.

Cryptantha grayi (Vasey & Rose) Macbr. var. *cryptochaeta* (Macbr.) I.M. Johnst. Local annual.

Heliotropium angiospermum Murray. Heliotrope, cabeza de violín, laven co'va (M). Widespread perennial herb.

Heliotropium fruticosum L. Local perennial herb.

Heliotropium macrostachyum (DC.) Hemsl. Heliotrope. Common subshrub.

Heliotropium procumbens Mill. Gray heliotrope. Common perennial herb.

Tournefortia hartwegiana Steudel. Confiturilla negra, tatachinole (M). Widespread shrub.

BUDDLEJACEAE

Buddleja sessiliflora H.B.K. Butterfly bush. Common shrub.

BURSERACEAE

Bursera fagaroides (H.B.K.) Engl. var. *elongata* McVaugh & Rzed. Torote de venado, torote de vaca, to'oro saguali (M). Common tree.

Bursera grandifolia (Schlect.) Engl. Palo mulato, to'oro mulato (M). Common tree.

Bursera lancifolia (Schlect.) Engl. [*B. fragilis* S. Watson]. Torote copal, to'oro chutama (M). Common tree.

*Bursera laxiflora S. Watson. Torote prieto, to'oro chucuri (M). Common tree.

Bursera penicillata (Sessé & Moc. ex DC.) Engl. Torote acensio [= de incienso], torote puntagruesa. Local tree.

Bursera stenophylla Sprague & Riley. Torote copal, to'oro sajo (M). Local tree.

CACTACEAE

Ferocactus pottsii (Salm.-Dyck) Backeb. Barrel cactus, biznaga, ónore (M). Local.

Mammillaria grahamii Engelm. Fishhook cactus, pitahayita, biznaguita, churrito, chicul ónore (M), tori bichu (M). Widespread.

Mammillaria standleyi (Britton & Rose) Orcutt. Biznaguita, pitahayita, bolsa de ratón. Common.

Opuntia pubescens Wendl ex Pfeiffer. Siviri chucha. Common.

Opuntia thurberi Engelm. Cholla, siviri. Widespread.

Opuntia wilcoxii Britton & Rose. Tree prickly pear, nopal, tuna, navo (M). Widespread.

Pachycereus pecten-aboriginum (Engelm.) Britton & Rose. Etcho (M). Widespread.

Pereskiopsis porteri (Weber) Britton & Rose. Jejari (M). Widespread.

Pilosocereus alensis Weber [*Cephalocereus alensis* (Weber) Britton & Rose]. Old man cactus, pitahaya barbona, aaqui jímsera (M). Local.

Selenicereus vagans (K. Brandegee) Britton & Rose. Sina volador, sina cuenojo (M). Local.

Stenocereus alamosensis (Coult.) Gibson & Horak [*Rathbunia alamosensis* Britton & Rose]. Galloping cactus, sina (M). Local.

Stenocereus montanus (Britton & Rose) F. Buxb. [*Lemairocereus montanus* Britton & Rose]. Sahuira (M). Local.

**Stenocereus thurberi* (Engelm.) Buxb. [*Lemairocereus thurberi* (Engelm.) Britton & Rose]. Organpipe cactus, pitahaya, aaqui (M). Widespread.

CAMPANULACEAE

Lobelia cordifolia Hook. & Arn. [*Heterotoma cordifolia* (Hook. & Arn.) McVaugh]. Common annual.

Lobelia fenestralis Cav. Local annual.

Lobelia goldmanii (Fern.) Ayers [*Heterotoma goldmanii* Fern.]. Local annual.

Triodanis biflora (Ruíz & Pav.) Greene. Venus looking glass. Local annual.

CANNABINACEAE

*Cannabis sativa L. Marijuana, marihuana, mota. Escaped illicit crop plant native to Old World. Local annual.

CAPPARACEAE

Cleome melanosperma S. Watson. Spider flower. Local annual.

Cleome viscosa L. Local annual.

CARICACEAE

Jarilla chocola Standl. Chócola (M). Local perennial herb.

CARYOPHYLLACEAE

Cerastium texanum Britton. Chickweed. Local annual.

Drymaria glandulosa C. Presl var. *glandulosa*. Cadenilla. Common annual.

Drymaria gracilis Cham. & Schlect. Cadenilla. Local annual.

CELASTRACEAE

Wimmeria mexicana (DC.) Lundell. Algodoncillo, chi'ini (M). Local tree.

CHENOPODIACEAE

Chenopodium album L. Goosefoot. Local annual.

Chenopodium ambrosioides L. Epazote, lipazote. Widespread perennial herb.

Chenopodium murale L. Introduced from Europe. Local annual.

Chenopodium neomexicanum Standl. Fishy goosefoot, quelite, chual (M). Local annual.

Chenopodium pratericola Rydb. Goosefoot, quelite, chual (M). Local annual.

COMPOSITAE (ASTERACEAE)

Acmella oppositifolia (Lam.) R.K. Jansen var. *oppositifolia*. Local perennial herb.

Ambrosia ambrosioides (Cav.) Payne. Canyon ragweed, chicura, jiogo (M). Widespread shrub.

Ambrosia confertiflora DC. Slimleaf bursage, estafiate, chíchibo (M). Widespread perennial herb.

Ambrosia cordifolia (A. Gray) Payne. Sonoran bursage, chicurilla, cau nachaqui (M). Widespread shrub.

Aster subulatus Michx. Common annual.

Baccharis salicifolia (Ruíz & Pav.) Pers. [*B. glutinosa* Pers.]. Seep willow, batamote, bacho'mo (M). Widespread shrub.

Bidens aurea (Ait.) Sherff. Common perennial herb.

Bidens bigelovii A. Gray. Local annual.

Bidens leptocephala Sherff. Beggar tick. Local annual.

Bidens pilosa L. Local annual.

Bidens riparia H.B.K. var. *refracta* O.E. Schulz. Saitilla. Local annual.

Bidens sambucifolia Cav. Widespread perennial herb or shrub.

Blumea viscosa (Mill.) Badillo. Local annual.

Brickellia coulteri A. Gray var. *adenopoda* (B.L. Rob.) B.L. Turner. Brickell bush, miona blanca, baco tami (M). Common subshrub.

Brickellia diffusa (Vahl) A. Gray. Widespread annual.

Brickellia sonorana B.L. Turner. Local annual.

Calea urticifolia (Mill.) DC. Local shrub.

Carminatia tenuiflora DC. Plume weed. Local annual.

Chaetymenia peduncularis Hook. & Arn. Common perennial herb.

Chloracantha spinosa (Benth.) Nesom var. *jaliscensis* (McVaugh) Sundberg [*Aster spinosus* Benth.]. Water aster, mo'oso (M). Widespread subshrub.

Conyza canadensis (L.) Cronq. Local annual.

Coreocarpus arizonicus (A. Gray) Blake. Local perennial herb.

Cosmos sulfureus Cav. Tostones. Common annual.

Delilia biflora (L.) Kuntze. Common annual.

Dyssodia anomala (Canby & Rose) B.L. Rob. Manzanilla coyota, manzanilla de coyote, goy manzanilla. Common annual.

Eclipta prostrata (L.) L. Mart. Hierba de tajo. Widespread annual.

Elephantopus spicatus Juss. Local perennial herb.

Erigeron velutipes Hook. & Arn. Fleabane. Common annual.

Eupatorium odoratum L. Boneset. Local shrub.

Eupatorium palmeri A. Gray var. *palmeri*. Boneset. Common shrub.

Eupatorium quadrangulare DC. Tree boneset, cocolmeca grande, lengua de buey, tepozana. Common subshrub.

Eupatorium surutatoanum B.L. Turner. Local subshrub.

Galinsoga parviflora Cav. Local annual.

Gamochaeta purpurea (L.) Cabrera. Annual cudweed. Common annual.

Guardiola platyphylla A. Gray. Local subshrub.

Helenium thurberi A. Gray. Common annual.

Heterotheca subaxillaris (Lam.) Britton & Rusby. Camphorweed, telegraph plant. Local perennial herb.

Hymenoclea monogyra Torr. & A. Gray. Cheesebush, romerillo, jeco (M). Widespread shrub.

Lagascea decipiens Hemsl. var. *glandulosa* (Fern.) Steussy. Confiturilla amarilla. Common shrub.

Lasianthaea fruticosa (L.) Becker var. *alamosana* Becker. Common shrub.

Machaeranthera tagetina Greene [*Aster tagetinus* (Greene) Blake]. Local perennial herb.

Melampodium appendiculatum B.L. Rob. Rosamarilla, manzanilla. Common annual.

Melampodium cupulatum A. Gray. Manzanilla. Local annual.

Melampodium perfoliatum (Cav.) H.B.K. Manzanilla. Local annual.

Milleria quinqueflora L. Common annual.

Montanoa leucantha (Lag.) Blake. subsp. *arborescens* V. A. Funk. Local shrub.

Montanoa rosei B.L. Rob. & Greenm. Batayaqui (M). Widespread shrub.

Parthenice mollis A. Gray. Local annual.

Parthenium hysterophorus L. Widespread annual.

Pectis coulteri Harv. & A. Gray. Goy sisi (M). Local annual.

Pectis prostrata Cav. Local annual.

Pectis uniaristata DC. Local annual.

Perityle californica Benth. Manzanilla. Common annual.

Perityle microglossa Benth. Widespread annual.

Pluchea carolinensis (Jacq.) G. Don. Local shrub.

Pluchea salicifolia (Mill.) Blake. Batamote. Local shrub.

Porophyllum coloratum (H.B.K.) DC. Local subshrub.

Porophyllum gracile Benth. Hierba de venado, cuchu pusi (M). Local subshrub.

Porophyllum macrocephalum DC. Odora. Common annual.

Pseudognaphalium leucocephalum (A. Gray) Anderb. Gordolobo, talampacate (M). Common perennial herb.

Sclerocarpus spatulatus Rose. Local annual.

Simsia setosa Blake. Local perennial herb.

Sonchus oleraceus L. Sow thistle. Introduced from Europe. Local annual.

Stevia ovata Willd. Local perennial herb.

Tagetes filifolia Lag. Local annual.

Tithonia fruticosa Canby & Rose. Girasol, mirasol. Local shrub.

Tridax procumbens L. Widespread perennial herb.

Tridax tenuifolia Rose. Local annual.

Trixis californica Kell. Local shrub.

Verbesina encelioides (Cav.) Benth. var. *encelioides*. Yellowtop. Local annual.

Vernonia triflosculosa H.B.K. var. *palmeri* (Rose) B.L. Turner. Local shrub.

Viguiera dentata (Cav.) Spreng var. *dentata*. Goldeneye, girasol. Local shrub.

Viguiera dentata (Cav.) Spreng var. *lancifolia* Blake. Goldeneye, santa maria, girasol. Common shrub.

Xanthium strumarium L. Cocklebur. Widespread annual.

Zinnia angustifolia H.B.K. var. *angustifolia*. Local perennial herb.

Zinnia zinnioides (H.B.K.) Olorode & Torres. Common annual.

CONVOLVULACEAE

Cuscuta americana L. Golden dodder. Parasite on *Haematoxylon* and *Senna pallida*. Local annual.

Cuscuta boldinghii Urban. Dodder. Parasite on *Dicliptera resupinata* and *Sida*. Local annual.

Cuscuta potosina Schaffner var. *globifera* Shaffner. Dodder. Parasite on *Ayenia filiformis* and *Evolvulus alsinoides*. Local annual.

Evolvulus alsinoides L. Common perennial herb.

Ipomoea arborescens (Humb. & Bonpl.) G. Don var. *glabrata* (A. Gray) Gentry. Morning glory tree, palo santo, cuta tósali (M), jútuguo (M). Common tree.

Ipomoea bracteata Cav. [*Exogonium bracteatum* (Cav.) G. Don.]. Jícama del monte, jícama (M), tosa huira (M). Common woody vine.

Ipomoea costellata Torr. Morning glory, trompillo. Local annual vine.

Ipomoea cristulata Hallier f. [*I. coccinea* L.]. Scarlet morning glory, trompillo. Common annual vine.

Ipomoea hederacea Jacq. Purple morning glory, trompillo. Local annual vine.

Ipomoea meyeri (Spreng.) G. Don. Pink morning glory, trompillo. Common annual vine.

Ipomoea minutiflora (Mart. & Gal.) House. Yellow morning glory, trompillo. Local annual vine.

Ipomoea pedicellaris Benth. Purple morning glory, trompillo, jícure (M). Local perennial vine.

Ipomoea quamoclit L. Scarlet morning glory, trompillo. Common annual vine.

Ipomoea triloba L. Morning Glory, trompillo. Common annual vine.

Ipomoea turbinata Lag. Spiny morning glory, trompillo. Local annual vine.

Jacquemontia azurea (Desr.) Choisy. Blue morning glory, trompillo. Local annual(?) vine.

Jacquemontia polyantha (Schlect. & Cham.) H. Hallier. White morning glory, trompillo. Local perennial vine.

Jacquemontia pringlei A. Gray. White morning glory, trompillo. Local perennial vine.

Merremia palmeri Hallier f. White morning glory, trompillo. Local perennial vine.

Merremia quinquefolia (L.) Hallier f. White morning glory, trompillo. Local perennial vine.

Operculina pteripes (G. Don) O'Donnell. Scarlet morning glory, trompillo. Local perennial vine.

CRUCIFERAE (BRASSICACEAE)

Coronopus didymus (L.) Smith. Wart or swine cress. Introduced from Eurasia. Local annual.

Descurainia pinnata (Walt.) Britton subsp. *halictorum* (Cockerell) Detling. Tansy mustard, pamita. Common annual.

Lepidium densiflorum Schrader. Lentejilla, peppergrass. Local annual.

Lepidium lasiocarpum Nutt. Lentejilla, peppergrass. Widespread annual.

Raphanus sativus L. Wild radish, rábano. Introduced from Europe. Local annual.

Rorippa teres (Michx.) Stuckey var. *rollinsii* Stuckey. Local annual.

Sisymbrium irio L. London rocket, pamitón. Introduced from Europe. Local annual.

CUCURBITACEAE

Citrullus lanatus (Thunb.) Mansf. [*C. vulgaris* Schrader]. Watermelon, sandía. Escaped cultivar native to tropical and southern Africa. Local annual vine.

Cucumis melo L. Melon de coyote, goy minole (M), hayá huíchibo (M). Escaped cultivar native to Near East. Local annual vine.

Cucurbita argyrosperma Huber subsp. *argyrosperma* Huber var. *palmeri* (L.H. Bailey) Merrick & Bates. Gourd, chichicayota, chichicoyota, tetaraca (M), hayá huíchibo (M). Widespread perennial vine.

Cyclanthera cf. *micrantha* Cogn. Local annual vine.

Echinopepon cirrhopedunculatus Rose. Common annual vine.

Ibervillea fusiformis (E.J. Lott) Kearns. Local perennial vine.

Lagenaria siceraria (Molina) Standl. Bottle gourd. Escaped cultivar. Local herbaceous vine.

Melothria pendula L. Common annual vine.

Schizocarpum palmeri Cogn. & Rose. Bitachera. Common annual vine.

Sechiopsis triquetra (Ser.) Naudine. Local annual vine.

Sicyos sinaloae Brandegee. Local annual vine.

Sicyosperma gracile A. Gray. Local annual vine.

EBENACEAE

Diospyros sonorae Standl. Sonoran persimmon, guayparín, caguorara (M). Common tree.

ERYTHROXYLACEAE

Erythroxylon mexicanum H.B.K. Mamoa, momoa, pañalío, momo ogua (M). Common shrub.

EUPHORBIACEAE

Acalypha aliena Brandegee. Local annual.

Acalypha cincta Müll. Arg. [*A. gentryi* Standl.]. Local shrub.

Acalypha ostryiifolia Riddell. Local annual.

Acalypha polystachya Jacq. Local annual.

Acalypha pseudalopecuroides Pax & K. Hoffm. Huevos de conejo, taguo bicho (M). Common annual.

Acalypha subviscida S. Watson. Common perennial herb.

Bernardia viridis Millsp. Common shrub.

Croton alamosanus Rose. Vara prieta, vara churi (M). Local shrub.

Croton ciliatoglandulifer Ortega. Ortiga, ortiguilla, trucha, rama blanca, tatio (M). Local shrub.

Croton fantzianus Seymour. Vara blanca, cuta tósari (M). Common tree.

Croton flavescens Greenm. Vara prieta, júsairo (M). Local shrub.

Dalechampia scandens L. Common perennial vine.

Ditaxis manzanilloana (Rose) Pax & K. Hoffm. [*Argythamnia gentryi* Ingram]. Local perennial herb.

Euphorbia alatocaulis V.W. Steinm. & Felger. Spurge, golondrina. Local perennial herb.

Euphorbia albomarginata Torr. & A. Gray. Golondrina. Local perennial herb.

Euphorbia capitellata Engelm. Spurge, golondrina, cuépari (M). Local perennial herb.

Euphorbia colletioides Bcnth. Candelillo. Widespread shrub.

Euphorbia cyathophora Murr. Poinsettia. Local annual.

Euphorbia densiflora (Klotzch & Garcke) Klotzch. Spurge, golondrina, cuépari (M). Common perennial herb.

Euphorbia dioscoreoides Boiss. subsp. *attenuata* V.W. Steinm. Local annual.

Euphorbia florida Engelm. Spurge, golondrina. Local annual.

Euphorbia gracillima S. Watson. Spurge, golondrina. Local annual.

Euphorbia graminea Jacq. Common annual.

Euphorbia heterophylla L. Common annual.

Euphorbia hirta L. Spurge, golondrina, cuépari (M). Widespread annual.

Euphorbia hyssopifolia L. Hyssop spurge, golondrina, cuépari (M). Widespread annual.

Euphorbia ocymoidea L. Spurge, golondrina, cuépari (M). Local annual.

Euphorbia petrina S. Watson. Spurge, golondrina, cuépari (M). Local annual.

Euphorbia serpens H.B.K. Spurge, golondrina, cuépari (M). Local annual.

Euphorbia sonorae Rose. Common annual.

Euphorbia thymifolia L. Spurge, golondrina, cuépari (M). Common perennial herb.

Jatropha cardiophylla (Torr.) Müll. Arg. Limberbush, sangrengado azul, sa'apo (M). Local shrub.

Jatropha cordata (Ortega) Müll. Arg. Torote papelío, torote amarillo, torote panalero, to'oro (M). Widespread tree.

Jatropha malacophylla Standl. [*J. platanifolia* Standl.]. Sangrengado, sa'apo (M). Common shrub.

Manihot aesculifolia (H.B.K.) Pohl. Local shrub.

Manihot angustiloba (Torr.) Müll. Arg. Local perennial herb.

Ricinus communis L. Castor bean, higuerilla. Introduced from Old World tropics. Widespread shrub.

Sebastiania pavoniana (Müll. Arg.) Müll. Arg. Mexican jumping bean, brincador, túbucti (M). Common tree.

Tragia sp. Vine noseburn, rama quemadora, ortiga, natari (M). Local perennial herb.

FAGACEAE

Quercus tuberculata Liebm. Encino amargo, bellota de cochi. Local tree.

FOUQUIERIACEAE

Fouquieria macdougalii Nash. Tree ocotillo, torote verde, jaboncillo, palo pitillo, ocotillo macho, murue (M). Common tree.

GENTIANACEAE

Centaurium nudicaule Engelm. Local annual.

HYDROPHYLLACEAE

Eucrypta chrysanthemifolia (Benth.) Greene. Local annual.

Annotated Checklist of Río Cuchujaqui Plants (continued)

Hydrolea spinosa L. var. *maior* (Brand) Davenp. Cilantro, chicalote. Common subshrub.

Nama hispidum A. Gray. Sand bells. Common annual.

Nama jamaicense L. Common annual.

Nama undulatum H.B.K. Local annual.

Phacelia gentryi Const. Sonoran caterpillai weed. Common annual.

Phacelia scariosa Brandegee. Local annual.

Wigandia urens (Ruíz & Pav.) var. *urens*. Quemadora. Local shrub.

KRAMERIACEAE

Krameria erecta Willd. [*K. parvifolia* Benth.]. Range ratany, tajui (M), tajimsi (M). Local shrub.

Krameria sonorae Britton. Sonoran ratany, cósahui (M). Local shrub.

LABIATAE (LAMIACEAE)

Hyptis albida Kunth. [*H. emoryi* Torr.]. Desert lavender, salvia, bíbino (M). Common shrub.

Hyptis mutabilis (Rich.) Briq. Common perennial herb.

Hyptis stellulata Benth. Local shrub.

Hyptis suaveolens (L.) Poit. Confiturilla, chan, comba'ari (M). Widespread annual.

Leonotis nepetaefolia A. Br. Lion's ears, cordoncillo de San Francisco. Native to South Africa. Common annual.

Salvia lasiocephala Hook. & Arn. Blue sage. Common annual.

Salvia mexicana L. Local subshrub.

Salvia misella Kunth. Local perennial herb.

Salvia riparia H.B.K. Local perennial herb.

Salvia setosa Fern. Sage. Widespread annual.

Stachys coccinea Jacq. Texas betony. Local perennial herb.

Teucrium cubense Jacq. subsp. *depressum* (Small) McClintock & Epling. Germander. Local annual.

LEGUMINOSAE (FABACEAE)

Acacia cochliacantha Humb. & Bonpl. ex Willd. Boatthorn acacia, güinolo, chírahui, chírajo (M). Widespread shrub.

Acacia farnesiana (L.) Willd. [*A. smallii* Isely]. Sweet acacia, vinorama, cu'uca (M). Common shrub.

Acacia occidentalis Rose. Catclaw, teso (M). Local shrub.

Aeschynomene fascicularis Schlect. & Cham. Local perennial herb.

Aeschynomene villosa Poir. var. *villosa*. Common annual.

Albizia sinaloensis Britton & Rose. Palo joso, joso (M). Common tree.

Astragalus sp. Locoweed. Local perennial herb.

Brongniartia alamosana Rydb. Palo piojo negro, cuta nahuila (M). Common tree.

Brongniartia nudiflora S. Watson [*B. palmeri* Rose]. Local shrub.

Caesalpinia palmeri S. Watson. Palo piojo, jícamuchi (M). Local shrub.

Caesalpinia platyloba S. Watson. Palo colorado, cuta síquiri (M), mapáo (M). Local tree.

Caesalpinia pulcherrima (L.) Sw. Red bird-of-paradise, tabachín, ta'boaca (M). Widely cultivated and naturalized ornamental in tropics of both hemispheres; area of origin unknown. Widespread shrub.

Calliandra houstoniana (Mill.) Standl. Local shrub.

Calliandra humilis Benth. var. *gentryana* Barneby. Fairy duster. Local shrub.

Canavalia brasiliensis Benth. Local woody vine.

Centrosema virginianum (L.) Benth. Local herbaceous perennial vine.

Cercidium praecox (S. Watson) Sarg. Brea, choy (M). Local tree.

Chamaecrista nictitans (L.) Moench. var. *mensalis* (Greenm.) Irwin & Barneby [*Cassia leptadenia* Greenm.]. Local annual.

Chloroleucon mangense (Jacq.) Britton & Rose var. *leucospermum* (Brandegee) Barneby & J.W. Grimes [*Pithecellobium undulatum* (Britton & Rose) Gentry]. Palo fierro, palo pinto, uose suctu (M), yoco suptu (M). Common tree.

Coursetia caribaea (Jacq.) Lavin var. *caribaea*. Common subshrub.

Coursetia glandulosa A. Gray. Samo (prieto), sa'mo (M), causamo (M). Common shrub.

Crotalaria cajanifolia H.B.K. Rattle box. Local shrub.

Crotalaria incana L. Rattle box. Local annual.

Crotalaria pumila Ortega. Rattle box. Local annual.

Crotalaria rotundifolia Gmel. var. *vulgaris* Windler. Rattle box. Local perennial herb.

Dalea cliffortiana Willd. Local annual.

Dalea elata Hook. & Arn. Common perennial herb.

Dalea exserta (Rydb.) Gentry. Local annual.

Dalea pringlei A. Gray var. *multijuga* Barneby. Popote. Local perennial herb.

Dalea scandens (Mill.) R.T. Clausen var. *occidentalis* (Rydb.) Barneby. Escoba, dáis, jíchiquia (M), yore cotéteca (M). Local shrub.

Desmanthus bicornutus S. Watson. Escoba, dáis, siteporo, jíchiquia (M). Local subshrub.

Desmodium angustifolium (H.B.K.) DC. var. *angustifolium*. Local perennial herb.

Desmodium prehensile Schldtl. Tick clover. Local perennial herb.

Desmodium procumbens (Mill.) Hitchc. var. *exiguum* (A. Gray) Schubert. Tick clover. Local annual.

Desmodium procumbens (Mill.) Hitchc. var. *longipes* (Schindler) Schubert. Tick clover. Local annual.

Desmodium procumbens (Mill.) Hitchc. var. *procumbens*. Tick clover. Local annual.

Desmodium psilocarpum A. Gray. Tick clover. Local annual.

Desmodium scopulorum S. Watson. Tick clover. Local annual or perennial herb.

Desmodium scorpiurus (Sw.) Desv. Sabino tick clover, frijol de godorniz (= codorniz), subuai muni (M). Widespread perennial herb.

Desmodium tortuosum (Sw.) DC. Local annual.

Annotated Checklist of Río Cuchujaqui Plants (continued)

Diphysa racemosa Rose. Árbol del diablo. Local shrub.

Erythrina flabelliformis Kearney. Coral bean, flor de Mayo, chilicote, pionía (fruit), jévero (M). Widespread tree.

Eysenhardtia orthocarpa (A. Gray) S. Watson. Kidneywood, palo dulce, baijguo (M). Local shrub.

Galactia cf. *wrightii* A. Gray. Local perennial vine.

Galactia sp. Common perennial vine.

Haematoxylum brasiletto Karst. Brasil, júchajco (M). Widespread tree.

Haematoxylum sp. nov. Brasil chino, tebcho (M). Local shrub.

Havardia mexicana (Rose) Britton & Rose [*Pithecellobium mexicanum* Rose]. Palo chino, chino (M). Local tree.

Havardia sonorae (S. Watson) Britton & Rose [*Pithecellobium sonorae* S. Watson]. Jócono, jócona (M). Local tree.

Indigofera jamaicensis Spreng. Local subshrub.

Indigofera suffruticosa Mill. Indigo, añil, mata de platanitos, chiju (M). Common shrub.

Leucaena lanceolata S. Watson. Guaje, palo bofo, güique (M). Local tree.

Leucaena leucocephala (Lam.) de Wit. Guaje. Introduced from tropical America. Local tree.

Lonchocarpus hermannii Marco [*Willardia mexicana* (S. Watson) Rose]. Nesco (M). Common tree.

Lysiloma divaricatum (Jacq.) J.F. Macbr. [*L. microphyllum* Benth.]. Mauto, mayo (M). Widespread tree.

Lysiloma watsonii Rose. Feather tree, tepeguaje, mach'agua (M). Common tree.

Marina alamosana (Rydb.) Barneby. Local perennial herb.

Marina diffusa (Moricand) Barneby. Local shrub.

Marina palmeri (Rose) Barneby. Rama de escoba. Local annual.

Marina scopa Barneby. Local annual.

Melilotus indicus (L.) All. Sour clover, trebolín. Introduced from Eurasia. Local annual.

Mimosa diplotricha C. Wright var. *odibilis* Barneby. Local annual.

Mimosa distachya Cav. var. *laxiflora* (Benth.) Barneby. Uña de gato, nésuquera (M). Local shrub.

Mimosa moniliformis (Britton & Rose) R. Gether & Barneby. Local shrub.

Mimosa palmeri Rose. Chopo, cho'opo (M). Common shrub.

Nissolia gentryi Rudd. Local woody vine.

Nissolia schottii A. Gray. Common woody vine.

Parkinsonia aculeata L. Mexican palo verde, retama, guacaporo (M), bacaporo (M). Widespread tree.

Phaseolus leptostachyus Benth. var. *leptostachyus*. Local annual.

Phaseolus lunatus L. Wild lima bean, frijolito, frijol de ratón. Local annual.

Pithecellobium dulce (Roxb.) Benth. Guamúchil, maco'tchini (M). Widespread tree.

Platymiscium trifoliolatum Benth. Tampicerán (M). Local tree.

Prosopis glandulosa Torr. var. *torreyana* (Benson) I.M. Johnst. Honey mesquite, mezquite, juupa (M). Widespread tree.

Rhynchosia minima (L.) DC. Frijolillo. Local annual vine.

Rhynchosia precatoria (Hook. & Benth. ex Willd.) DC. Rosary bean, chana (chanate) pusi (M). Common woody vine.

Senna atomaria (L.) Irwin & Barneby [*Cassia emarginata* L.]. Palo zorrillo, jupachumi (M). Widespread tree.

Senna hirsuta (L.) Irwin & Barneby var. *glaberrima* (M.E. Jones) Irwin & Barneby [*Cassia leptocarpa* Benth.]. Slimpod senna. Local perennial herb.

Senna obtusifolia (L.) Irwin & Barneby. Ejotillo. Local annual.

Senna occidentalis (L.) Link. Ejotillo, ejotillo cafecillo. Local annual.

Senna pallida (Vahl) Irwin & Barneby var. *shreveana* Irwin & Barneby [*Cassia biflora* L.]. Garbancillo, flor de iguana, huicori bísaro (M). Common shrub.

Senna uniflora (Mill.) Irwin & Barneby. Ejotillo chiquito. Local annual.

Sesbania herbacea (Mill.) McVaugh. Baiquillo (M). Widespread annual.

Sphinctospermum constrictum (S. Watson) Rose. Local annual.

Tephrosia rhodantha Brandegee. Local perennial herb.

Tephrosia cf. *vicioides* Sch. Local perennial herb.

Zapoteca formosa (Kunth) H. Hern. subsp. *rosei* (Wiggins) H. Hern. [*Calliandra formosa* Kunth.]. Local shrub.

Zornia reticulata J.E. Smith [*Z. diphylla* (L.) Pers.]. Hierba de la víbora. Local perennial herb.

LENNOACEAE

*Pholisma culiacanum (Dressler & Kuijt) Yatskievych [*Ammobroma culiacana* Dressler & Kuijt] (S.A. Meyer, obs., Oct. 1991). Hongo. Local parasite.

LOASACEAE

Eucnide hypomalaca Standl. Rock nettle, ma'aco arócosi (M). Local perennial herb.

Gronovia scandens L. Widespread perennial vine.

Mentzelia aspera L. Stickleaf, buena mujer. Common annual.

LOGANIACEAE

Polypremum procumbens L. Local annual.

LORANTHACEAE

Struthanthus palmeri Kuijt. Mistletoe, toji (M), chíchel (M). Common parasite on *Brongniartia alamosana, Guazuma ulmifolia, Prosopis glandulosa, Melia azedarach, Salix bonplandiana*, and *Taxodium distichum*.

LYTHRACEAE

Cuphea laminuligera Koehne in Mart. Common annual.

Cuphea llavea Lex. in La Llave & Lex. Local subshrub.

Cuphea lobophora Koehne var. *occidentalis* S.A. Graham. Local annual.

Rotala ramosior (L.) Koehne. Tooth cup. Local annual.

Annotated Checklist of Río Cuchujaqui Plants (continued)

MALPIGHIACEAE

Callaeum macropterum (DC.) D.M. Johnson [*Mascagnia macroptera* (Moc. & Sessé) Ndzu]. Gallinitas, batanene, matanene, sana rogua (M). Widespread woody vine.

Echinopterys eglandulosa (Adr. Juss.) Small. Local woody vine.

Heteropterys palmeri Rose. Compio (M). Local woody vine.

Janusia californica Benth. Desert vine. Local woody vine.

Malpighia emarginata DC. [*M. umbellata* Rose]. Granadilla, morita, sire (M). Local tree.

MALVACEAE

Abutilon abutiloides (Jacq.) Britton & Wilson. Indian mallow, malva, pintapán, ta'ari sorogua (M). Common shrub.

Abutilon mollicomum (Willd.) Sweet [*A. sonorae* A. Gray]. Indian mallow. Local subshrub.

Abutilon mucronatum J. Fryxell. Indian mallow, malva, rama escoba, to'oro cojuya (M). Widespread subshrub.

Abutilon reventum S. Watson. Indian mallow, malva colorada. Common subshrub.

Abutilon trisulcatum (Jacq.) Urb. Indian mallow, malva, rama escoba, to'oro cojuya (M). Local subshrub.

Allosidastrum hilairianum (C. Presl) Krapovickas, Fryxell & Bates. Local shrub.

Anoda abutiloides A. Gray. Local subshrub.

Anoda acerifolia Cav. Local perennial herb.

Anoda crenatiflora Ortega. Local annual.

Anoda cristata (L.) Schlect. Local annual.

Anoda lanceolata Hook. & Arn. Local perennial herb.

Anoda reticulata S. Watson. Local annual.

Anoda thurberi A. Gray. Local perennial herb.

Bastardiastrum cinctum (Brandegee) Bates. Malva blanca, escoba. Common shrub.

Herissantia crispa (L.) Brizicky. Malva, malva negra, chicuri. Widespread subshrub.

Hibiscus biseptus S. Watson. Rose mallow. Widespread perennial herb.

Hibiscus citrinus Fryxell. Rose mallow. Local shrub.

Malvastrum bicuspidatum (S. Watson) Rose subsp. *bicuspidatum*. Malva peluda, tuchi. Common shrub.

Malvastrum coromandelianum (L.) Garcke. Local perennial herb.

Sida abutifolia Mill. Local perennial herb.

Sida aggregata C. Presl. Malva blanca, rama lisita, sayla de pintapán, to'oro jíchiquia (M). Local shrub.

Sida ciliaris L. Local perennial herb.

Sida glabra Mill. Local annual.

Sida hyalina Fryxell. Local subshrub.

Sida linifolia Cav. Local perennial herb.

Sida rhombifolia L. Local subshrub.

Sphaeralcea coulteri (S. Watson) A. Gray. Globe mallow. Local annual.

Wissadula hernandioides L'Herit. Local annual.

MARTYNIACEAE

Martynia annua L. Aguaro (M), tan cócochi (M). Local annual.

Proboscidea altheaefolia (Benth.) Decne. Devil's claw, aguaro, perrito. Local perennial herb.

Proboscidea parviflora (Woot.) Woot. & Standl. subsp. *parviflora*. Devil's claw, perrito, aguaro (M). Local annual.

MELIACEAE

Trichilia americana (Sessé & Moc.) Pennington [*T. colimana* C. DC.]. Piocha, bola colorada, síquiri tájcara (M). Local tree.

MENISPERMACEAE

Cocculus diversifolius DC. Snail seed, huichuri. Common woody vine.

MOLLUGINACEAE

*Glinus radiatus (Ruíz & Pav.) Rohrb. Local annual.

Mollugo verticillata L. Indian chick weed, carpet weed. Native of temperate South America. Local annual.

MORACEAE

Dorstenia drakeana L. Baiburilla, pionía, na'atoria (M). Common perennial herb.

Ficus cotinifolia H.B.K. Strangler fig, higuera, nacapuli (M). Common tree.

*Ficus insipida Willd. [*F. radulina* S. Watson]. Fig, chalate, higuera, chuna, tchuna (M). Local tree.

Ficus pertusa L. f. [*F. padifolia* H.B.K.]. Fig, camuchín, capulín, chuna, nacapuli (M), tchuna (M). Local tree.

Ficus petiolaris H.B.K. subsp. *petiolaris*. Rock fig, tescalama, báisaguo (M). Common tree.

*Ficus trigonata L. [*F. goldmanii* Standl.]. Fig, chalate colorado, higuera. Local tree.

MYRTACEAE

Psidium guajava L. Guayabo, guava, cogoyo (M). Introduced cultivated tree native to southern Mexico. Local tree.

Psidium sartorianum (Berg.) Ndzu. Arrellán. Local tree.

NYCTAGINACEAE

Boerhavia diffusa L. [*B. coccinea* Mill.]. Red spiderling, mochis (M), juana huipili (M). Widespread perennial herb.

Boerhavia erecta L. var. *erecta*. Spiderling, mochis (M). Local annual.

Boerhavia gracillima Heimerl. Red spiderling, mochis. Local perennial herb.

Boerhavia spicata Choisy. Spiderling, mochis. Common annual.

Boerhavia xanti S. Watson. Spiderling, mochis. Local annual.

Commicarpus scandens (L.) Standl. Miona (M). Common subshrub.

Pisonia capitata (S. Watson) Standl. Garambullo, baijuo (M). Widespread woody vine.

Salpianthus arenarius Humb. & Bonpl. [*S. macrodontus* Standl.]. Guayavilla, juéoguo (M). Common perennial herb.

Salpianthus purpurascens (Lag.) Hook. & Arn. Jarilla. Common perennial herb.

Annotated Checklist of Río Cuchujaqui Plants (continued)

OLACACEAE

Schoepfia schreberi J.F. Gmel. Palo cachora, cuta béjori (M), júchica (M). Common tree.

ONAGRACEAE

Gaura parviflora Dougl. Lizard tail. Local annual.

Ludwigia octovalvis (Jacq.) Raven var. *octovalvis*. Water primrose. Widespread perennial herb.

Ludwigia peploides (H.B.K.) Raven. Yellow water weed, verdolaga de agua. Common aquatic.

Oenothera kunthiana (Spach.) Munz. Common perennial herb.

OPILIACEAE

Agonandra racemosa (DC.) Standl. Palo verde, matachamaco, úsim yuera (M). Common tree.

OXALIDACEAE

Oxalis corniculata L. Wood sorrel, chanchaquilla. Common perennial herb.

PAPAVERACEAE

Argemone ochroleuca Sweet subsp. *ochroleuca*. Mexican prickly poppy, cardo santo, chicalote, táchino (M). Widespread annual.

Papaver somniferum L. Opium poppy, amapola. Escaped illicit crop plant native to Eurasia. Local annual.

PASSIFLORACEAE

Passiflora foetida L. var. *gossypiifolia* (Desv.) Masters. Passionflower, bule de venado, huirote de venado, talayote, ma'aso alócossim (M), ma'aso huiroa (M). Local perennial vine.

Passiflora suberosa L. Passionflower. Local perennial vine.

PEDALIACEAE

Sesamum indicum L. Sesame, ajonjolí. Escaped crop plant native to southern Europe. Local annual.

PHYTOLACCACEAE

Petiveria alliacea L. Zorrillo, jupachumi (M). Common perennial herb.

Phaulothamnus spinescens A. Gray. Ba'co (M). Local shrub.

Rivina humilis L. Pigeonberry, coralito cimarrón. Local perennial herb.

PIPERACEAE

Piper jaliscanum S. Watson. Cocolmeca. Local shrub.

PLATANACEAE

Platanus racemosa Nutt. var. *wrightii* (S. Watson) L. Benson. Arizona sycamore, aliso. Local tree.

PLUMBAGINACEAE

Plumbago scandens L. Hierba de alacrán, pitillo, cresta de gallo, toto'ora (M). Common perennial herb.

PODOSTEMACEAE

Oserya cf. *coulteriana* Tul. Rock water moss. Local aquatic.

POLEMONIACEAE

Bonplandia geminiflora Cav. Ciática (M). Common perennial herb.

Ipomopsis sonorae (Rose) A. Grant [E. Palmer 396 in 1890 (US)]. Local annual.

Loeselia ciliata L. Acacia. Local perennial herb.

Loeselia glandulosa (Cav.) Don. Acacia, trigo, tiriscu (M). Common perennial herb.

Loeselia pumila (Mart. & Gal.) Walp. Local perennial herb.

POLYGALACEAE

Polygala glochidiata H.B.K. Milkwort. Local annual.

Polygala aff. *obscura* Benth. Milkwort. Local perennial herb.

POLYGONACEAE

Antigonon leptopus Hook. & Arn. Sanmiguelito, masasari (M). Common perennial vine.

Polygonum persicaria L. Lady's thumb, moco de guajolote. Introduced from Europe. Common perennial herb.

Rumex inconspicuus Rech. f. Dock. Common perennial herb.

PORTULACACEAE

Portulaca oleracea L. Purslane, verdolaga, guaro (M). Widespread annual.

Portulaca pilosa L. Local perennial herb.

Talinum paniculatum (Jacq.) Gaertn. Pink baby breath, rama del sapo. Local perennial herb.

Talinum triangulare (Jacq.) Willd. Common perennial herb.

PRIMULACEAE

Anagallis arvensis L. Pimpernel, hierba de pájaro. Introduced from Europe. Local annual.

RANUNCULACEAE

Clematis drummondii Torr. & A. Gray. Virgin's bower, barba de chivato. Local woody vine.

RHAMNACEAE

Colubrina triflora Brongn. Cuta béjori (M). Local shrub.

Gouania rosei Wiggins. Huirote de violín. Widespread woody vine.

Karwinskia humboldtiana (Roem. & Sch.) Zucc. Cacachila, aroyoguo (M). Common shrub.

*Ziziphus amole (Sessé & Moc.) M.C. Johnst. [*Z. sonorensis* S. Watson]. Saituna, báis cápora (M). Local tree.

RUBIACEAE

Cephalanthus occidentalis L. var. *salicifolius* (Humb. & Bonpl.) A. Gray. Button bush, mimbro. Widespread shrub.

Chiococca petrina Wiggins. Local shrub.

Crusea coronata B.L. Rob. & Greenm. Local annual.

Crusea psyllioides (H.B.K.) Anderson. Local annual.

Crusea setosa (Mart. & Gal.) Standl. & Steyerm. Local annual.

Annotated Checklist of Río Cuchujaqui Plants (continued)

Diodia sarmentosa Sw. Common perennial vine.

Galium proliferum Ait. Bedstraw. Local annual.

Hedyotis vegrandis W.H. Lewis. Local annual.

Hintonia latiflora (Sessé & Moc.) Bullock [*Coutarea latiflora* Sessé & Moc.].
    Copalquín, tapichogua (M). Common tree.

Mitracarpus hirtus (L.) DC. Common annual.

Randia echinocarpa Sessé & Moc. ex DC. Papache, jósoina (M). Widespread shrub.

Randia laevigata Standl. (S.A. Meyer obs.). Sapuchi (M). Local shrub.

Randia obcordata S. Watson. Papache borracho, pi'isi (M). Common shrub.

Randia thurberi S. Watson. Papache borracho, pi'isi (M). Local shrub.

Richardia scabra L. Local perennial herb.

Spermacoce confusa Rendle. Local annual.

RUTACEAE

Casimiroa edulis La Llave & Lex. White sapote, chapote, zapote, ja'apa ahuim (M).
    Escaped cultivated tree native from Sinaloa to Jalisco. Local tree.

Esenbeckia hartmani B.L. Rob. & Fernald. Palo amarillo, jójona (M), momoguo (M).
    Local shrub.

Zanthoxylum fagara (L.) Sarg. Gato, limoncillo, matalased, o'ou sesutu (M). Local
    shrub.

Zanthoxylum mazatlanum Sandwith. Gato. Local shrub.

SALICACEAE

Populus mexicana Sarg. subsp. *dimorpha* (Brandegee) Eckenwalder. Mexican
    cottonwood, álamo, aba'aso (M). Planted by I. Acosta; washed away by
    September 1995 floods of Hurricane Ismael. Local tree.

Salix bonplandiana H.B.K. Bonpland willow, sauce, huata (M). Common tree.

Salix gooddingii Ball. Goodding willow, sauce, huata (M). Local tree.

Salix taxifolia H.B.K. Yewleaf willow, sauce. Local shrub.

SAPINDACEAE

Cardiospermum corindum L. Balloon vine, bombita. Local woody vine.

Cardiospermum cuchujaquense M.S. Ferrucci & Acev.-Rodr. Cuchujaqui balloon vine.
    Local woody vine.

Sapindus saponaria L. Soapberry tree, amolillo, abolillo, tubchi (M). Widespread tree.

Serjania mexicana Willd. Enredadera de culebra, huirote de culebra, camugia (M).
    Widespread woody vine.

Serjania palmeri S. Watson. Huirote de cachora, camugia (M). Local woody vine.

SAPOTACEAE

Sideroxylon persimile (Hemsl.) Pennington subsp. *subsessiliflorum* (Hemsl.)
    Pennington [*Bumelia arborescens* Rose]. Bebelama (M). Local tree.

Sideroxylon tepicense (Standl.) Pennington [*S. angustifolium* Standl.]. Tempisque,
    ca'ja (M). Common tree.

SCROPHULARIACEAE

Antirrhinum costatum Wiggins. Snapdragon. Common annual.

Bacopa repens (Sw.) Wettst. Disc water hyssop. Local aquatic.

Castilleja tenuiflora Benth. Indian paint brush. Local perennial herb.

Leucospora intermedia (A. Gray) Keil var. *intermedia* [*Schistophragma intermedia* (A. Gray) Pennell]. Local annual.

Leucospora intermedia (A. Gray) Keil var. *subglabra* Keil. Local annual.

Mecardonia vandellioides (H.B.K.) Pennell. Widespread perennial herb.

Mimulus cardinalis Benth. Cardinal monkey flower. Local perennial herb.

Mimulus floribundus Dougl. Little yellow monkey flower. Common annual.

Mimulus guttatus DC. Yellow monkey flower. Common perennial herb.

Russelia sonorensis Carlson var. *pubescens* Carlson. Common subshrub.

Stemodia durantifolia (L.) Sw. var. *durantifolia*. Widespread perennial herb.

Stemodia palmeri A. Gray. Widespread annual.

Veronica peregrina L. subsp. *xalapensis* (H.B.K.) Pennell. Necklace weed. Common annual.

SIMAROUBACEAE

Alvaradoa amorphoides Liebm. Palo torsal, guaji (M), paenepa (M). Local shrub or tree.

SOLANACEAE

Callibrachoa parviflora (Juss.) D'Arcy [*Petunia parviflora* Juss.]. Wild petunia. Common annual.

Capsicum annuum L. var. *aviculare* (Dierbach) D'Arcy & Eshbaugh. Chiltepín, có'ocori (M). Local subshrub.

Datura discolor Bernh. Desert thorn apple, toloache, tebue (M). Common annual.

Datura lanosa Bye. Sacred datura, toloache, tebue (M). Common perennial herb.

Lycium andersonii A. Gray. Wolfberry, sigropo (M). Local shrub.

Nicotiana glauca Graham. Tree tobacco, cornetón, ojo de coyote, tronadora. Introduced from South America. Local shrub.

Nicotiana obtusifolia Mart. & Gal. [*N. trigonophylla* Dunal]. Wild tobacco, tabaco de coyote, tabaquillo, goy biba (M). Local perennial herb.

Nicotiana plumbaginifolia Viviani. Wild tobacco. Common annual.

Physalis acutifolia (Miers) Sandwith. Ground cherry, tomatillo. Common annual.

Physalis hederaefolia A. Gray. Ground cherry, tomatillo. Local perennial herb.

Physalis leptophylla B.L. Rob. & Greenm. Shiny ground cherry, tombrisi (M). Local annual.

Physalis minima L. Ground cherry, tomatillo. Local annual.

Physalis philadelphica Lamarck. Tomatillo, tombrísi (M). Escaped cultivated food plant. Local annual.

Physalis pubescens L. Ground cherry. Local annual.

Physalis subulata Rydb. Ground cherry. Local annual.

Solanum adscendens Sendtler [*S. deflexum* Greenm.]. Nightshade, chichiquelite, mambia (M). Local annual.

Annotated Checklist of Río Cuchujaqui Plants (continued)

Solanum americanum Mill. [*S. nodiflorum* Jacq.]. Nightshade, chichiquelite, mambia (M). Local annual.

Solanum erianthum D. Don. [*S. verbascifolium* L.]. Cornetón del monte, higo de la sierra, sosa, lengua de buey, buey nini (M), chivera jimsi (M). Local shrub or tree.

Solanum ferrugineum Jacq. [*S. madrense* Fern.]. Sosa. Common shrub.

Solanum grayi Rose. Nightshade. Common annual.

Solanum lumholtzianum Bartlett. Local annual.

Solanum lycopersicum L. [*Lycopersicon esculentum* Mill.]. Tomato, tomate. Escaped cultivated food plant native to Andean South America. Common perennial herb.

Solanum nigrescens Mart. & Gal. [*S. douglasii* Dunal]. Nightshade, chichiquelite, mambia (M). Widespread perennial herb.

Solanum seaforthianum Andrews. Vine nightshade, bellissima. Escaped ornamental native to tropical America. Local woody vine.

Solanum tridynamum Dunal [*S. amazonium* Ker.]. Nightshade, saca manteca, ojo de liebre, poros pusi (M). Common shrub.

Solanum umbellatum Mill. Sosa. Local shrub.

STERCULIACEAE

Ayenia filiformis S. Watson. Common perennial herb.

Ayenia filipes (S. Watson) McVaugh. Local shrub.

Ayenia jaliscana S. Watson. Local shrub.

Ayenia mexicana Turcz. Local shrub.

Byttneria aculeata Jacq. Almohada de culebra, huira espinosa, baco muteca (M). Common woody vine.

Guazuma ulmifolia Lam. Guásima, agia (M). Widespread tree.

Melochia pyramidata L. Malva. Common perennial herb.

Melochia tomentella (C. Presl) Hemsl. Malva. Common subshrub.

Waltheria acuminata Rose. Local shrub.

Waltheria detonsa A. Gray. Common perennial herb.

Waltheria indica L. [*W. americana* L.]. Common perennial herb.

TAMARICACEAE

Tamarix ramosissima Ledeb. Tamarisk, pino salado. Introduced from Eurasia. Common shrub.

THEOPHRASTACEAE

Jacquinia macrocarpa Cav. subsp. *pungens* (A. Gray) B. Ståhl. Sanjuanico, tásiro (M). Widespread shrub.

TILIACEAE

Corchorus hirtus L. Orinoco jute. Local annual.

Corchorus siliquosus L. Local annual.

Heliocarpus attenuatus S. Watson. Samo baboso, sa'amo (M). Local tree.

Heliocarpus palmeri S. Watson. Samo prieto, algodoncillo. Local shrub.

Triumfetta semitriloba L. Guachapori de los cerros, guachaporito (M). Local subshrub.

ULMACEAE

Celtis iguanea (Jacq.) Sarg. Tropical hackberry, palo pinto, vainoro, cumbro (M). Widespread woody vine.

Celtis pallida Torr. Desert hackberry, cumbro (M). Local shrub or tree.

Celtis reticulata Torr. Netleaf hackberry, cúmero, aceituna, cumbro (M). Local tree.

UMBELLIFERAE (APIACEAE)

Coriandrum sativum L. Coriander, cilantro. Escaped cultivated spice plant native to Mediterranean region. Local annual.

Daucus pusillus Michx. Wild carrot. Common annual.

Eryngium nasturtiifolium Delar. f. Hierba del sapo. Widespread annual.

*Eryngium sp. Local perennial herb.

Hydrocotyle cf. *umbellata* L. Water pennywort. Local aquatic.

Spermolepis echinata (Nutt.) Heller. Scaleseed. Local annual.

URTICACEAE

Parietaria hespera D.B. Hinton. Pellitory. Local annual.

Pouzolzia palmeri S. Watson. Local shrub.

VERBENACEAE

Bouchea dissecta S. Watson. Local annual.

Glandularia gooddingii (Briq.) Solbrig [*Verbena gooddingii* Briq]. Goodding vervain. Local perennial herb.

Glandularia pumila (Rydb.) Umber [*Verbena pumila* Rydb.]. Local annual.

Lantana achyranthifolia Desf. Confiturilla blanca. Common subshrub.

Lantana camara L. [*L. horrida* H.B.K.]. Confiturilla negra, ta'ampisa (M). Common shrub.

Lantana hispida H.B.K. [*L. velutina* Mart. & Gal.]. Confiturilla blanca. Common subshrub.

*Lippia cf. *graveolens* H.B.K. Orégano de burro, bura mariola. Local shrub.

Lippia umbellata Cav. Orégano. Local tree.

Phyla reptans (H.B.K.) E. Greene var. *reptans*. Diamondleaf frogfruit. Local perennial herb.

Priva lappulacea (L.) Pers. Cordoncillo. Common perennial herb.

Verbena officinalis L. Vervain. Common perennial herb.

Vitex mollis H.B.K. Uvalama, igualama, ugualama, júvare (M). Widespread shrub (tree).

VIOLACEAE

Hybanthus attenuatus (Humb. & Bonpl.) C.K. Schulze. Local annual.

VISCACEAE

Phoradendron quadrangulare (Kunth) Krug & Urban [*P. guazumae* Trel.]. Parasite on *Salix bonplandiana*. Mistletoe, toji (M). Local perennial.

Annotated Checklist of Río Cuchujaqui Plants (continued)

VITACEAE

Cissus verticillata (L.) Nicholson & Jarvis. Huirote de agua, yucohuira (M). Widespread woody vine.

ZYGOPHYLLACEAE

Guaiacum coulteri A. Gray. Guayacán, júyaguo (M). Common tree.

Kallstroemia parviflora Norton. Summer poppy, baiburín, jímuri (M). Local annual.

MONOCOTS

AGAVACEAE

**Agave vilmoriniana* Berger [*A. mayoensis* Gentry]. Octopus agave, amole. Local succulent.

Agave vivipera L. Maguey, mezcal del monte, babki, juya cu'u (M). Widespread succulent.

ALISMATACEAE

Echinodorus berteroi (Spreng.) Fassett var. *berteroi*. Burhead, hierba del manso del agua, patito. Common aquatic.

Sagittaria montevidensis Cham. & Schlect. Arrow head, flecha de agua. Local aquatic.

ARACEAE

Xanthosoma hoffmannii (Schott) Schott. Polongo. Local perennial herb.

BROMELIACEAE

Hechtia montana Brandegee. False agave, mezcalito, aguamita (M). Local succulent.

Tillandsia elizabethae Rauh [*T. sonorensis* L.B. Smith]. Local perennial herb.

Tillandsia recurvata (L.) L. Ball moss, mezcalito de árbol, mezcalito de huitacochi, huíribis cúu (M), pipisqui (M). Uncommon epiphyte on many plants and substrates. Local.

COMMELINACEAE

Callisia monandra (Sw.) Schult. Teléfono cimarrón. Local annual.

Commelina cf. *caroliniana* Walt. Dayflower, hierba del pollo. Local perennial herb.

Commelina diffusa Burm. f. Widespread perennial herb.

Commelina erecta L. Dayflower, cañaigra, hierba del pollo, cuna del niño, osi (M). Flowers both blue and white. Local perennial herb.

Tinantia longipedunculata Standey & Ste. Local annual.

Tradescantia cf. *crassifolia* Cav. Spider flower. Local perennial herb.

Tripogandra palmeri (Rose) Woodson. Local annual.

CYPERACEAE

Cyperus compressus L. Flatsedge, coquillo, coni saquera (M). Local perennial.

Cyperus dentoniae G.C. Tucker. Flatsedge, coquillo. Common perennial.

Cyperus difformis L. Flatsedge, coquillo. Local annual.

Cyperus entrerianus Boeck. Flatsedge, coquillo. Local perennial.

Cyperus esculentus L. Yellow nutgrass, coquillo. Introduced from Southeast Asia. Local perennial.

Cyperus fugax Liebm. Flatsedge, coquillo. Local annual.

Cyperus involucratus Rottb. Flatsedge, coquillo. Introduced; type collection from Arabia. Local perennial.

Cyperus iria L. Flatsedge, coquillo. Introduced from Southeast Asia. Common annual.

Cyperus ochraceus Vahl. Flatsedge, coquillo. Common perennial.

Cyperus odoratus L. Flatsedge, coquillo. Local perennial.

Cyperus pallidicolor (Kük.) G.C. Tucker. Flatsedge, coquillo. Local perennial.

Cyperus rotundus L. Purple nutsedge. Introduced from Eurasia. Local perennial.

Cyperus squarrosus L. Flatsedge, coquillo. Local annual.

Cyperus surinamensis Rottb. Flatsedge, coquillo. Local perennial.

Cyperus tenerrimus J. Presl & C. Presl. Flatsedge, coquillo. Local perennial.

Eleocharis atropurpurea (Retz.) Kunth. Spike rush, zacatito del agua. Local annual.

Eleocharis geniculata (L.) Roem. & Schult. Spike rush. Local annual.

Eleocharis montana (H.B.K.) Roem. & Schult. var. *nodulosa* (Roth) Svenson. Spike rush. Local perennial.

Fimbristylis dichotoma (L.) Vahl. Local annual.

Lipocarpha micrantha (Vahl) G.C. Tucker. Common annual.

DIOSCOREACEAE

Dioscorea convolvulacea Schlect. & Cham. Local woody vine.

GRAMINEAE (POACEAE)

Aegopogon cenchroides Humb. & Bonpl. ex Willd. Local annual.

Aristida adscensionis L. Six-weeks threeawn. Local annual.

Aristida ternipes Cav. var. *ternipes*. Spider grass, ba'asso (M). Widespread perennial.

Arundinella palmeri Beal. Cañuela, carricillo. Local perennial.

Arundo donax L. Giant cane, carrizo, ba'ca (M), ba'aca nagua (M). Introduced from Old World. Common perennial.

Axonopus compressus (Sw.) P. Beauv. Local perennial.

Bothriochloa barbinodis (Lag.) Herter. Cane beardgrass, popotillo. Local perennial.

Bothriochloa pertusa (L.) A. Camus. Bluestem. Introduced from Old World. Local perennial.

Bouteloua alamosana Vasey. Alamos grama, navajita. Local annual.

Bouteloua aristidoides (H.B.K.) Griseb. Six-weeks needle grama, aceitilla, zacate saitilla, yemsa ba'asso (M). Common annual.

Bouteloua barbata Lag. var. *barbata*. Six-weeks grama, navajita, ma'as ba'asso (M). Local annual.

Bouteloua chondrosioides (H.B.K.) S. Watson. Sprucetop grama, navajita. Local perennial.

Bouteloua curtipendula (Michx.) Torr. Sideoats grama, banderilla. Local perennial.

Bouteloua quiriegoensis Beetle. Quiriego grama, navajita. Local annual.

Bouteloua radicosa (E. Fourn.) Griffith. Purple grama, navajita. Local perennial.

Bouteloua repens (H.B.K.) Scribn. & Merr. Slender grama, navajita delgada. Common perennial.

Brachiaria arizonica (Scribn. & Merr.) S.T. Blake. Piojillo. Local annual.

Annotated Checklist of Río Cuchujaqui Plants (continued)

Brachiaria fasciculata (Sw.) L. Parodi. Piojillo, granadillo. Common annual.

Brachiaria reptans (L.) C.A. Gard. & C.E. Hubb. Local annual.

Cathestecum brevifolium Swallen. False grama, grama china, zacate borreguero. Widespread perennial.

Cenchrus brownii Roem. & Schult. Sandbur, guachapori, na'a chú'uqui (M). Local annual.

Cenchrus echinatus L. Southern sandbur, guachapori, na'a chú'uqui (M). Local annual.

Cenchrus incertus M.A. Curtis. Field sandbur, guachapori, na'a chú'uqui (M). Local annual.

Chloris chloridea (J. Presl) Hitchc. Fingergrass, chinhuirinhuis. Local perennial.

Chloris virgata Sw. Feather fingergrass, zacate lagunero, cola de zorra. Common annual.

Cynodon dactylon (L.) Pers. Bermuda grass, zacate de lana. Introduced from tropical-subtropical Africa. Widespread perennial.

Dactyloctenium aegyptium (L.) P. Beauv. Crowfoot grass, pata de grulla. Introduced from Old World. Common annual; weak perennial.

Dichanthium annulatum (Forssk.) Stapf [*Andropogon annulatus* Forssk.]. Kleberg bluestem, fuzzynode. Introduced from tropical-subtropical Africa to India and China. Local perennial.

Dichanthium aristatum (Poir.) C.E. Hubb. [*Andropogon aristatus* Poir.]. Angleton bluestem. Introduced from subtropical Asia. Local perennial.

Diectomis fastigiata (Sw.) P. Beauv. [*Andropogon fastigiatus* Sw.]. Pasto mota. Local annual.

Digitaria bicornis (Lam.) Roem. & Schult. Zacate dos cuernos, huilanchi (M). Local annual.

Digitaria ciliaris (Retz.) Koeler. Southern crabgrass. Common annual.

Digitaria horizontalis Willd. Local annual.

Echinochloa colonum (L.) Link. Jungle rice, zacate pinto, zacate lagunero. Introduced from Old World. Common annual.

Echinochloa crusgallii (L.) P. Beauv. Barnyard grass, zacate pinto rayado, zacate de corral. Introduced from Old World. Local annual.

Eleusine indica (L.) Gaertn. Goose grass. Introduced from Eurasia. Common annual.

Eragrostis cilianensis (All.) Janchen. Stink grass. Introduced from Europe. Common annual.

Eragrostis ciliaris (L.) R. Br. Common annual.

Eragrostis intermedia Hitchc. Plains lovegrass. Local perennial.

Eragrostis lugens Nees. Lovegrass, zacate llorón cimarrón, zacate llanero tosco. Local annual.

Eragrostis maypurensis (H.B.K.) Steud. Lovegrass. Local annual.

Eragrostis mexicana (Hornem.) Link subsp. *mexicana*. Mexican lovegrass. Local annual.

Eragrostis pectinacea (Michx.) Nees var. *miserrima* (E. Fourn.) J. Reeder. Lovegrass. Local annual.

Eragrostis pectinacea (Michx.) Nees var. *pectinacea*. Lovegrass. Widespread annual.

Eriochloa aristata Vasey. Cup grass. Common annual.

Gouinia virgata (J. Presl) Scribn. Surguillo, zacate colorado, ba'asso síquiri (M). Local perennial.

Hackelochloa granularis (L.) Kuntze. Zacate granillo. Local annual.

Heteropogon contortus (L.) Roem. & Schult. Tanglehead, zacate colorado. Common perennial.

Lasiacis ruscifolia (H.B.K.) Hitchc. Negrito, carricito de la sierra, ba'ca caupojomi (M). Widespread perennial.

Leptochloa mucronata (Michx.) Kunth [*L. filiformis* (Lam.) P. Beauv.]. Six-weeks sprangletop. Widespread annual.

Leptochloa panicoides (J. Presl) Hitchc. Widespread annual or perennial.

Leptochloa virgata (L.) P. Beauv. Local annual.

Muhlenbergia arizonica Scribn. Arizona muhly, liendrilla, zacate tarango. Local perennial.

Muhlenbergia dumosa Vasey. Bamboo muhly, otatillo. Local perennial.

Muhlenbergia elongata Beal. Cliff muhly, liendrilla. Widespread perennial.

Muhlenbergia fragilis Swallen. Liendrilla. Local annual.

Muhlenbergia microsperma (DC.) Trinius. Six-weeks muhly, liendrilla. Widespread annual.

Muhlenbergia rigens (Benth.) Hitchc. Deer grass. Local perennial.

Muhlenbergia tenella (H.B.K.) Trinius. Rock hair, liendrilla. Common annual.

Oplismenus burmannii (Retz.) P. Beauv. Zacate cadillo. Local annual.

Oplismenus hirtellus (L.) P. Beauv. Zacate barbón. Local annual.

Panicum hians Elliott. Gaping panicgrass. Local perennial.

Panicum hirticaule J. Presl var. *verrucosum* Zuloaga & Morrone. Witchgrass, panizo guarito, zacate capiro. Local annual.

Panicum pampinosum Hitchc. & Chase. Local annual.

Panicum trichoides Sw. Common annual.

Panicum virgatum L. Switchgrass. Local perennial.

Paspalum convexum Humb. & Bonpl. Local perennial.

Paspalum distichum L. Knotgrass. Zacate de arena. Local perennial.

Paspalum hartwegianum E. Fourn. Local perennial.

Paspalum humboldtianum Flügge. Local perennial.

Paspalum langei (E. Fourn.) Nash. Local perennial.

Paspalum paniculatum L. Local perennial.

Paspalum squamulatum E. Fourn. Local perennial.

Pennisetum ciliare (L.) Link [*Cenchrus ciliaris* L.]. Buffelgrass, zacate buffel. Introduced from eastern Africa. Widespread perennial.

Annotated Checklist of Río Cuchujaqui Plants (continued)

Phragmites australis (Cav.) Steud. Carrizo. Local perennial.

Polypogon monspeliensis (L.) Desf. Rabbitfoot grass, cola de zorra. Introduced from Europe. Local annual.

Rhynchelytrum repens (Willd.) C.E. Hubb. Natal grass, zacate rosado. Introduced from South Africa. Widespread perennial.

Setaria liebmannii E. Fourn. Tropical six weeks bristlegrass, cola de zorra, hayás guasia (M). Common annual.

Setaria parviflora (Poir.) Kerguélen. Arroyo bristlegrass, tempranero gusanillo. Common perennial.

Setariopsis auriculata (E. Fourn.) Scribn. Plumerillo, zacate orejón. Local annual.

Sorghum halepense (L.) Pers. Johnson grass, zacate Johnson. Introduced from Old World. Common perennial.

Tripsacum lanceolatum E. Fourn. Mexican gamagrass, milpa de venado, zacate maíz. Local perennial.

JUNCACEAE
Juncus tenuis Willd. Slender rush. Local perennial.

LEMNACEAE
Lemna aequinoctialis Welwitsch. Lama, berenogua (M). Local aquatic.

Lemna minuscula Herter. Duckweed. Lama, berenogua (M). Local aquatic.

LILIACEAE
Aloe barbadensis Mill. [*A. vera* L.]. Aloe vera, sábila. Local African succulent established on roadside.

Echeandia ramosissima (C. Presl) Cruden. White crag lily. Local perennial herb.

Hymenocallis sonorensis Standl. Spider lily, lirio de los arroyos. Widespread perennial herb.

LIMNOCHARITACEAE
Limnocharis flava (L.) Buch. Flowering rush. Local aquatic.

NAJADACEAE
Najas guadalupensis (Spreng.) Magnus. Common water nymph. Local aquatic.

ORCHIDACEAE
Encyclia adenocarpon (Lex.) Schltr. [*Epidendron adenocarpa* Lex.]. Local perennial herb.

Habenaria quinqueseta (Michx.) Eaton. Spider orchid. Local perennial herb.

Oncidium cebolleta (Jacq.) Sw. Cuerno de chivita. Local perennial herb.

Spiranthes polyantha Reichb. f. Local perennial herb.

PALMAE (ARECACEAE)
Brahea aculeata (Brandegee) Moore [*Erythea aculeata* Brandegee]. Palmilla, palma, ta'aco (M). Common tree.

Sabal uresana Trel. Palma, palma real. Local tree.

PONTEDERIACEAE
*Eichhornia crassipes (Mart.) Solms-Laubach. Water hyacinth. Local floating

perennial aquatic herb native to South America. Unsuccessfully introduced at the Güirocoba crossing in 1993.

Heteranthera limosa (Sw.) Willd. Mud plaintain. Local aquatic.

POTAMOGETONACEAE

Potamogeton foliosus Raf. Pondweed. Local aquatic.

Potamogeton nodosus Poir. Pondweed. Local aquatic.

TYPHACEAE

Typha domingensis Pers. Cattail, tule. Common perennial.

ZANNICHELLIACEAE

Zannichellia palustris L. Common poolmat. Local aquatic.

REFERENCES

Beetle, A. A., and J. G. Johnson. 1991. *Gramineas de Sonora*. Hermosillo: Secretaría de Agricultura y Recursos Hidráulicos.

Brown, D. E., and C. H. Lowe. 1980. *Biotic communities of the Southwest*. U.S. Forest Service General Technical Report RM-78 (map). Fort Collins, Colo.: Rocky Mountain Forest and Range Experimental Station.

Brown, D. E., C. H. Lowe, and C. P. Pase. 1979. A digitized classification system for the biotic communities of North America, with community (series) and association examples for the Southwest. *Journal of the Arizona-Nevada Academy of Science* 14:1–16.

Bullock, S. H. 1995. Plant reproduction in neotropical dry forests. In *Seasonally dry tropical forests*, eds. S. H. Bullock, H. A. Mooney, and E. Medina, 277–303. Cambridge: Cambridge University Press.

Bullock, S. H., H. A. Mooney, and E. Medina, eds. 1995. *Seasonally dry tropical forests*. Cambridge: Cambridge University Press.

Bye, R. A., Jr. 1995. Ethnobotany of the Mexican tropical dry forests. In *Seasonally dry tropical forests*, eds. S. H. Bullock, H. A. Mooney, and E. Medina, 423–438. Cambridge: Cambridge University Press.

Cuevas, E. 1995. Biology of the belowground system of tropical dry forests. In *Seasonally dry tropical forests*, eds. S. H. Bullock, H. A. Mooney, and E. Medina, 362–383. Cambridge: Cambridge University Press.

Felger, R. S. 1980. Vegetation and flora of the Gran Desierto, Sonora, Mexico. *Desert Plants* 2:87–114.

Ferrucci, M. S., and P. Acevedo-Rodríguez. 1998. *Cardiospermum cuchujaquense* (Sapindaceae), a new species from Sonora, Mexico. *Novon* 8:235–238.

French, R. 1962. Alamos—Sonora's city of silver. *Smoke Signal* 5:1–16.

Friedman, S. L. 1996. Vegetation and flora of the coastal plains of the Río Mayo region, southern Sonora, Mexico. Masters thesis, Arizona State University, Tempe.

Gentry, A. H. 1995. Diversity and floristic composition of neotropical dry forests. In *Seasonally dry tropical forests*, eds. S. H. Bullock, H. A. Mooney, and E. Medina, 146–194. Cambridge: Cambridge University Press.

Gentry, H. S. 1942. *Río Mayo plants: A study of the flora and vegetation of the valley of the Río*

Mayo, Sonora. Publication No. 527. Washington, D.C.: Carnegie Institution of Washington.

————. 1982. Sinaloan deciduous forest. *Desert Plants* 4:73–77.

Holbrook, N. M., J. L. Whitbeck, and H. A. Mooney. 1995. Drought responses of neotropical dry forest trees. In *Seasonally dry tropical forests*, eds. S. H. Bullock, H. A. Mooney, and E. Medina, 243–276. Cambridge: Cambridge University Press.

Kearney, T. H., and R. H. Peebles. 1969. *Arizona flora*. 2d ed. with suppl., eds. J. T. Howell and E. McClintock. Berkeley: University of California Press.

Krizman, R. D. 1972. Environment and season in a tropical deciduous forest in northwestern Mexico. Ph.D. dissertation, University of Arizona, Tucson.

Lenz, L. W. 1992. *Plants of the Cape region, Baja California Sur, Mexico*. Claremont, Calif.: Cape Press.

Lott, E. J. 1993. *Annotated checklist of the vascular flora of the Chamela Bay region, Jalisco, Mexico*. Occasional Papers of the California Academy of Sciences No. 148.

Martin, P. S., D. A. Yetman, M. E. Fishbein, P. D. Jenkins, T. R. Van Devender, and R. Wilson, eds. 1998. *Gentry's Río Mayo plants: The tropical deciduous forest and environs of northwest Mexico*. Tucson: University of Arizona Press.

Martínez-Y., A. 1995. Biomass distribution and primary productivity of tropical dry forests. In *Seasonally dry tropical forests*, eds. S. H. Bullock, H. A. Mooney, and E. Medina, 326–345. Cambridge: Cambridge University Press.

McVaugh, R. 1983. *Flora Novo-Galiciana. Volume 14: Gramineae*. Ann Arbor: University of Michigan Press.

————. 1984. *Flora Novo-Galiciana. Volume 12: Compositae*. Ann Arbor: University of Michigan Press.

————. 1987. *Flora Novo-Galiciana. Volume 5: Leguminosae*. Ann Arbor: University of Michigan Press.

Medina, E. 1995. Diversity of life forms of higher plants in neotropical dry forests. In *Seasonally dry tropical forests*, eds. S. H. Bullock, H. A. Mooney, and E. Medina, 221–243. Cambridge: Cambridge University Press.

Reichenbacher, F. W., S. Franson, and D. E. Brown. 1998. *The biotic communities of North America*. Salt Lake City: University of Utah Press.

Rondeau, R. J., T. R. Van Devender, C. D. Bertelsen, P. Jenkins, R. K. Wilson, and M. A. Dimmitt. 1996. Annotated flora and vegetation of the Tucson Mountains, Pima County, Arizona. *Desert Plants* 12:3–46.

Rose, J. N. 1891. List of plants collected by Dr. Edward Palmer in 1890 in western Mexico and Arizona at Alamos. *Contributions to the United States National Herbarium* 1:91–116.

Rzedowski, J. 1978. *Vegetación de México*. México, D.F.: Editorial Limusa.é

Rzedowski, J., and R. McVaugh. 1966. La vegetación de Nueva Galicia. *Contributions from the University of Michigan Herbarium* 9:1–123.

Shelford, V. E. 1963. *The ecology of North America*. Urbana: University of Illinois Press.

Shreve, F., and I. L. Wiggins. 1964. *Vegetation and flora of the Sonoran Desert*. Stanford: Stanford University Press.

Spellenberg, R., T. Lebgue, and R. Corral-Díaz. 1996. *A specimen-based, annotated checklist of the vascular plants of Parque Nacional "Cascada de Basaseachi" and adjacent areas, Chihuahua, Mexico*. Listados Florísticos de México No. 12.

Standley, P. C. 1920–1926. Trees and shrubs of Mexico. *Contributions to the United States National Herbarium* 23:1–1721.

Steinmann, V. W. 1996. A revision of the *Euphorbia dioscoreoides* complex (Euphorbiaceae). *Aliso* 14:219–226.

Steinmann, V. W., and R. S. Felger. 1997. The Euphorbiaceae of Sonora, Mexico. *Aliso* 16:1–71.

Toolin, L. J., T. R. Van Devender, and J. M. Kaiser. 1979. The flora of Sycamore Canyon, Pajarito Mountains, Santa Cruz County, Arizona. *Journal of the Arizona-Nevada Academy of Sciences* 14:66–74.

Van Devender, T. R. 1995. Desert grassland history: Changing climates, evolution, biogeography, and community dynamics. In *The desert grassland*, eds. M. P. McClaran and T. R. Van Devender, 68–99. Tucson: University of Arizona Press.

Van Devender, T. R., C. H. Lowe, and H. E. Lawler. 1994. Factors influencing the distribution of the Neotropical vine snake *(Oxybelis aeneus)* in Arizona and Sonora, Mexico. *Herpetological Natural History* 2:27–44.

Van Devender, T. R., A. C. Sanders, V. W. Steinmann, R. K. Van Devender, S. A. Meyer, J. F. Wiens, D. A. Yetman, P. D. Jenkins, E. López-S, R. A. López-E, and J. D. Freeh. 1995. Noteworthy collections: Sonora. *Madroño* 42:411–418.

Wiggins, I. L. 1980. *Flora of Baja California*. Stanford: Stanford University Press.

Yetman, D. A., T. R. Van Devender, P. Jenkins, and M. Fishbein. 1995. The Río Mayo: A history of studies. *Journal of the Southwest* 37:294–345.

4

MONTE MOJINO

Mayo People and Trees in Southern Sonora

David A. Yetman, Thomas R. Van Devender,
Rigoberto A. López Estudillo, and
Ana Lilia Reina Guerrero

Scattered across the southern portion of Sonora into the northern extremity of Sinaloa are a few dozen Mayo Indian villages, tiny enough that only a small number of persons outside the area have heard of them. These unassuming hamlets are sufficiently isolated that visits from the outside world are infrequent and the object of considerable local curiosity.

This paper presents information on the trees in tropical deciduous forest (TDF) and Sinaloan thornscrub and their uses, which we gleaned from our extensive interviews with Mayo consultants who live in these villages and settlements near Alamos, Sonora.

The Mayo

The Mayo language is classified as Cáhita, a group that includes Yaqui and numerous extinct languages. Mayo is sufficiently close to Yaqui that the two tongues are mutually intelligible (Camou 1985). Mayos inhabited the region long before the European conquest of Mexico. The first Europeans found them to be an agricultural and fishing people who were friendly to the advancing Spaniards, perhaps because they hoped to find allies in the militarily imposing forces from Spain for their intermittent wars with the Yaquis to the north (Spicer 1962). Unlike the Yaquis, the Mayos offered no organized resistance to the Spaniards or to the imposition of Catholicism. Indeed, testimony from Fr. Pedro Méndez, an early Jesuit missionary, paints in 1614 a picture of a docile and fervently religious people: "I have never seen in any converts so clear a sign of the grace and presence of the Holy Spirit as among these [Mayos]. Even those still set in old customs, on being baptized, appear

overcome with such extraordinary joy that the lame and aged seem to recover their feet and their youthful energy, and the mute recover their speech,
so that they can run to the church and to their priest and give thanks for his
merciful gift to them" (Acosta 1949).

By the early years of the seventeenth century, most Mayos had been baptized and welcomed to the fold of the converted. Those who resisted did
not survive. Over the next 250 years, Mayos rebelled periodically, objecting
to enslavement for work in the mines or on haciendas and at times reject-
ing the stringent religious regulations imposed upon them by the Jesuits.
After the expulsion of the Jesuits in 1767, the Mayos were subjected to extreme exploitation in the form of peonage on haciendas or as laborers in the
mines of Alamos. In the nineteenth century, they joined two organized re-
bellions, teaming with Yaquis in their hope to drive non-Mayos from their
lands. Lacking strong pan-Mayo organization, however, they were finally
defeated in the 1880s and have since showed signs of being absorbed into the
dominant Mexican culture. By the early 1900s, nearly all traditional Mayo
lands had been appropriated by Mexican *latifundistas*, the wealthy owners of
great landed estates (Spicer 1962). Since the Mexican revolution and especially since the administration of President Lázaro Cárdenas in the 1930s,
many villages have been constituted into *ejidos* (communally owned lands)
or *comunidades* (community-owned lands) under title from the Mexican government. Older Mayos view Cárdenas as the last Mexican president who
cared about *la gente campesina* (peasants), even though the lands they received are only a small percentage of their aboriginal territory.

Although Mayos are Sonora's largest indigenous group (Camou 1985),
their villages are typically small, seldom containing more than a few dozen
families. In the *municipios* (counties) of Huatabampo and Etchojoa, a few
larger Mayo communities are found with strong cultural traditions. In other
areas the villages are widely dispersed, communication among them is often
nonexistent (Crumrine 1983), and they are not identified as Mayo. In contrast, most Yaquis live in large towns recognized as Yaqui towns, with highly
organized community structures that encourage constant communication
among the Yaquis who still live in the Yaqui Indigenous Land (Spicer 1980).

Our research shows that contemporary Mayo populations are largely concentrated in three areas: (1) a small foothills group on and near the Río Mayo
north and east of Navojoa to the vicinity of San Bernardo, (2) a coastal group
located on the coastal plain in the municipios of Huatabampo and Etchojoa, which grades into (3) a Sinaloan group in the river drainages from the

Río Mayo south into northern Sinaloa. Communication among these three populations is sporadic at best and appears to be maintained through festivities associated with Lent and Holy Week, a period of intense participation by most Mayos. Some Sinaloan Mayos view the groups living on the Río Mayo as excessively influenced by Yaquis, whom they view as rogue sorcerers and violent, lawless plunderers. In the villages we studied, the Mayo language is used extensively among older individuals but seldom among younger Mayos and children. All of the Mayos we contacted spoke Spanish fluently, though often with an undeniable accent. Only in the coastal group does the Mayo tongue (referred to as *la lengua*) appear to be widely spoken by young people. Even in the strongly Mayo village of Yocogigua, children speak Spanish exclusively. The village is without *pascola* dancers, a deeply embedded Mayo tradition, the absence of which is a clear indicator of cultural change. Nevertheless, they refer to themselves, as do all Mayos, as *yoreme*; only outsiders use the term *Mayo* when referring to Mayos.

Study Areas

We selected three villages and a rancho within an 80-km radius of Alamos in southern Sonora because their rural orientation and location in or on the fringes of largely intact TDF permit them to retain familiarity with plants and their uses. Although we visited other sites, it was in the company of consultants from these three villages. The sites also represent a cross section of vegetation types of southern Sonora, ranging from coastal thornscrub on coastal flats, foothills thornscrub on rocky hillsides below 200-m elevation, riparian communities along a large arroyo, and TDF in the Sierra de Alamos and northward.

Teachive is a Sinaloan Mayo village at 75-m elevation in the municipio of Navojoa 40 km southeast of Navojoa, lying southwest of the Sierra de Alamos. The name means "scattered round stones" in the Mayo language. At Teachive, the Arroyo Masiaca, which originates in the Sierra de Alamos and the Sierra Batayaqui to the north, has permanent surface water. Some 24 km south, it empties directly into the Gulf of California at Las Bocas, originally a Mayo fishing village, now a semideveloped Mexican beach resort. To the west and south of Teachive, the foothills thornscrub merges into a highly varied coastal thornscrub dominated by pitahaya *(Stenocereus thurberi)*, in places forming spectacular cactus forests or *pitahayales* (fig. 4.1). We visited the following sites: thornscrub on the road to Las Bocas near Teachive and

FIGURE 4.1. Pitahayal (coastal thornscrub dominated by *Stenocereus thurberi*) 2.3 km northeast of Las Bocas, Sonora (September 1994). Photograph by Thomas R. Van Devender.

on the rocky slopes of Cerro Terúcuchi at an elevation of 120 m (3 km north of Teachive); riparian communities along the Arroyo Masiaca near Teachive and Las Bocas; and TDF at 240–460 m at Las Rastras on the southwest slopes of the Sierra de Alamos (about 20 km northwest of Teachive).

Teachive is part of the large Masiaca comunidad and is more affluent than other villages in the study area, owing to its favorable supply of water and agricultural land and the industry of its women, who produce woven wool *cobijas* (blankets) of high quality and great beauty. The weaving industry has perpetuated a familiarity with many plants in the region, which are used in one phase or another in the production of blankets. One man also carves ceremonial masks and dance figures from a variety of woods. In recent years several families have begun producing intricately woven wire baskets, an art which they have perfected.

Yocogigua, "where the jaguar eats" in Mayo, is a Sinaloan Mayo settlement of about 50 families in the municipio of Alamos, 19 km northeast of Teachive. The vegetation on the surrounding gentle foothills at 240 to 320 m is a transition between TDF and Sinaloan thornscrub. We visited sites just east and south of the village: Cerro El Chorro (4 km northwest) and the El Bariste road (10 km west). We also visited riparian forests with sabino (*Taxodium distichum*) at 180 m on the Río Cuchujaqui at San Vicente (12 km east).

In the last two decades, Yocogigua has experienced severe economic depression. A mescal factory, which had produced the fiery distillate for four decades and had employed most of the townspeople, was shut down in 1974 and now lies in a state of magnificent decay. The ejido of Yocogigua was created under the administration of President Echeverría in 1975. Although the ejido is large (6,800 ha) and possesses excellent forest resources, the local economy is limited to the cutting of firewood for sale in nearby Huatabampo and Navojoa, the raising of small numbers of livestock, and occasional day labor, such as working in the distant fields of the Río Mayo delta and harvesting the seed of buffelgrass *(Pennisetum ciliare)*. The rapid disappearance of usable firewood bodes ill for the future of the community and the strong Mayo culture that now exists there. The absence of employment has a demoralizing effect on the town, especially on the young men.

El Rincón Viejo (commonly called El Rincón) is a small Mayo rancho of four homes on the edge of the series of mountains that are the northern extensions of the Sierra de Alamos (5 km north of Alamos). Here a trail begins at about a 500-m elevation and rises to 720 m as it traverses the Cerro Piedra Boluda into Arroyo Las Bebelamas (560 m), all on Ejido Las Bebelamas. We also visited canyon riparian gallery forest along the Río Cuchujaqui at 350 m at Rancho El Guayabo (14 km east of Alamos).

El Rincón lies on the edge of the large ejido Las Bebelamas, of which our consultants are *socios* or members. The consultants from the settlement have a variety of sources of employment, although some of these are limited by government prohibitions on the cutting of wood.

Nahuibampo is a foothills Mayo village in the municipio of Alamos 50 km northwest of Alamos and 10 km west-southwest of San Bernardo. It is located approximately 80 km northwest of Yocogigua on a rolling terrace above the Río Mayo, which adds a riverine dimension to the vegetation. Here we visited TDF at an elevation of 300 to 400 m on *la falda*, the "skirt" of Cordón Topiyecas on ejido land 4 km north-northwest of the village.

Nahuibampo is a poor village indeed, and the Ejido Nahuibampo is a poor ejido, although it is maintained by its members with great dignity. Although it sits on the banks of the Río Mayo, irrigable land is absent and dryland farms are unreliable. Many cattle died in the terrible drought of 1993–1994 and again in 1995–1996, and the prospects for recovery are dim. Few youths are seen in the village; the rest have fled to the cities with the hope of finding work.

Vegetation

The thornscrub in the vicinity of Teachive is dominated by arborescent species reaching a height of roughly 6 m. Brasil *(Haematoxylum brasiletto)*; brea *(Cercidium praecox)*; cacachila *(Karwinskia humboldtiana)*; copalquín *(Hintonia latiflora)*; etcho *(Pachycereus pecten-aboriginum)*; granadilla *(Malpighia emarginata)*; guayacán *(Guaiacum coulteri)*; güinolo, chírahui, or boatthorn acacia *(Acacia cochliacantha)*; jaboncillo or tree ocotillo *(Fouquieria macdougalii)*; mezquite *(Prosopis glandulosa)*; palo blanco *(Piscidia mollis)*; palo de asta *(Cordia sonorae)*; palo fierro *(Chloroleucon mangense)*; torote papelío *(Jatropha cordata)*; saituna *(Ziziphus amole)*; sanjuanico *(Jacquinia macrocarpa)*; torote copal *(Bursera lancifolia)*; and torote prieto *(B. laxiflora)* are but some of the typical tree species. Along the arroyo margins are dense groves of guásima *(Guazuma ulmifolia)*, mezquite, and palo joso *(Albizia sinaloensis)*.

A series of small but dramatic basaltic hills 1 km to the north adds several other species, including agaves and cacti, to the list of flora. Like all Sonoran rangelands, the communal lands of the Masiaca community, of which Teachive is a part, are heavily grazed, posing a challenge to the plant collector to identify species before they are consumed or trampled by sheep, goats, and cattle.

In the vicinity of Yocogigua, only slightly higher in elevation but closer to the foothills of the Sierra de Alamos, the tree species are more numerous than nearer the coast. Canopy height exceeds 8 m or more. In addition to the species mentioned above, the following species are found: algodoncillo *(Wimmeria mexicana)*, amapa *(Tabebuia impetiginosa)*, mauto *(Lysiloma divaricatum)*, nesco *(Lonchocarpus hermannii)*, palo colorado *(Caesalpinia platyloba)*, palo mulato *(Bursera grandifolia)*, palo piojo *(Brongniartia alamosana)*, palo piojo blanco *(Caesalpinia caladenia)*, palo santo *(Ipomoea arborescens)*, palo zorrillo *(Senna atomaria)*, pochote *(Ceiba acuminata)*, and samo baboso *(Heliocarpus attenuatus)*. Palo joso is especially abundant along the ephemeral watercourse that meanders through the hamlet's center; its graceful shape, lacy leafage, and tall, elegant stature add an element of dignity to an already picturesque village.

The vegetation at El Rincón is greatly altered by generations of town dwellers, making comparison with the other sites difficult. Comparison of the Alamos area today with photographs from 50 years ago indicates that

FIGURE 4.2. Vara blanca *(Croton fantzianus)* poles along Alamos-Navojoa road west of El Caracol, ready for shipment to Baja California for use in grape and tomato fields (September 1994). Photograph by Thomas R. Van Devender.

general vegetation, including trees, is more abundant and varied than earlier, a pleasant effect of the cessation of mining in the region. In the arroyo to the northeast of El Rincón and on the northwest side of the range, including the moist canyon bottom, these species thrive: bebelama *(Sideroxylon persimile)*; chapote *(Casimiroa edulis)*; several species of figs *(Ficus cotinifolia, F. insipida*, and *F. petiolaris)*; guayavillo *(Acacia coulteri)*; matachamaco or palo verde *(Agonandra racemosa)*; palo barriga *(Cochlospermum vitifolium)*; palo bofo *(Leucaena lanceolata)*; sahuira *(Stenocereus montanus)*; sangrengado *(Jatropha malacophylla)*; and torote acensio *(Bursera penicillata)*. The ejido is lumbered extensively for vara blanca *(Croton fantzianus*; fig. 4.2), which is harvested in lengths somewhat longer than 2 m, hauled by burro to collection points, and sold in great quantities to buyers, who in turn market the poles to tomato and grape farmers in Baja California.

Nahuibampo, in the drainage of the Río Mayo, lies at the edge of a formerly vast, forested upland that extended nearly unbroken into the tortured canyons and escarpment of the Sierra Sutucamé to the north and west. Palo mulato, torote de vaca *(Bursera fagaroides)*, and torote prieto are common. Algodoncillo is prevalent, as is amapa. Other prominent trees include chilicote *(Erythrina flabelliformis)*, chopo *(Mimosa palmeri)*, mauto, tempisque *(Sideroxylon tepicense)*, and vara blanca, mixed in a remarkably open forest whose canopy exceeds 10 m. The rainfall at the 300- to 400-m elevation supports a more varied and taller forest than that at Teachive and Yocogigua but not as diverse as that at Las Rastras and El Rincón. Portions of the ejido are occasionally cleared for *milpas*, hand-cleared plots used for a few years to raise corn and beans and then allowed to revert to forest.

Climate

The Mayo region is virtually frost free. In 1937 Gentry (Gentry 1942) noted remnants of a freeze that occurred several years earlier. It damaged some trees but left most untouched. A frost struck some areas in December 1997, but the damage was spotty. Such events are truly rare. The growth cycle of plants in the region is determined by rainfall, not by cold temperatures. Temperatures in the early summer drought, on the other hand, sometimes exceed 40°c. The peak temperatures occur while most trees are leafless.

The bulk of the precipitation in the Mayo region falls during a 9-week period beginning in the first week of July and ending by the middle of September. These rains, *las aguas* as they are called locally, are typically heavy but brief. Lesser amounts of rain frequently fall during the winter. These more gentle storms, called *equipatas*, are usually less reliable than the summer showers, though they may last for several days at a time, testing the integrity of roofs and forcing a human closeness not experienced at other times. An elderly Mayo woman related that her mother had pronounced September 15 to be the date of the end of las aguas. Rains occurring after that date and through the winter were equipatas, she proclaimed.

From February through the end of June, rain is unusual. Five months often pass without moisture. During that period, especially in the searing hot months of May and June, most trees drop their leaves and the landscape takes on a brownish gray aspect, an appearance called *mojino* by Sonorans. Only the green of the etcho and pitahaya cacti interrupt the monotony of dull-colored plains and hillsides. In watercourses and canyons, however, the great figs and a few tropical trees retain their green aspect. Although outsiders find the late spring oppressive owing to stifling heat, pervasive dust, and air commonly laden with smoke from crop burning, the Mayos find it a time when the absence of dense understory herbs and climbers permits easier access to the *monte* (bush) and their fields and when the long, relatively cool mornings and low humidities afford pleasant working conditions.

A shorter, less certain drought occurs from mid-September to December.

During these months the smothering heat of summer gradually dissipates, and the long autumn, when the trees lose their leaves, begins. The sequence of deciduations is remarkable. Beginning within 10 days of the last drops of las aguas, the leaves of one species after another turn color, fade, and trickle to the ground. *Jatropha cordata* is the first to release its leaves, which turn yellow rapidly and fall noiselessly to the ground. Winds are infrequent and the forest is usually still, disturbed only by light and transient gusts. Those attuned to nature's subtleties can perceive the delicate drift of dried leaves sifting to the earth. On the higher slopes, the immense leaves of palo joso de la sierra *(Conzattia multiflora)* are the first to turn. The branches are rapidly emptied of their foliage, and within a few days the tree stands bare. The last leaves on guásima and palo zorrillo may endure until April. Those of güinolo may persevere through the drought.

Since 1990 substantial fall rains following dry to very dry summers have tended to retard the rate of deciduation in the forest. Trees that have begun to deciduate appear to halt their dropping of leaves, and others delay their deciduation. The forest takes on an unusual appearance under these circumstances. Trees that have not deciduated retain a strongly green aspect, whereas those whose leaves have fallen tend to be obscured by the greenery. The intense emerald of las aguas, however, is absent. If the trend toward delay of the onset of las aguas and the proliferation of autumnal rains continues, new dynamics will surely appear in the forest.

Several tree species, especially tepeguaje *(Lysiloma watsonii)*, seem to anticipate the rainy season and leaf out in advance of the first rain. Some early sprouting species may accomplish this by sensing the increased humidity that precedes the onset of summer storms, thus gaining an early growth advantage when the first drops arrive. In contrast, other species are extremely conservative. In the case of palo joso de la sierra, the failure of summer rains in 1993 resulted in a failure of some trees to leaf out.

Annual precipitation ranges from approximately 275 mm on the coastal plain to approximately 350 mm at Teachive to about 610 mm at San Bernardo, a few kilometers from Nahuibampo (Camou 1985; Hastings and Humphrey 1969). At Las Rastras and El Rincón on the edges of the Sierra de Alamos, the rich vegetation suggests rainfall considerably higher than at the other locations. Rainfall varies greatly in most years. During las aguas in 1994, a station in Alamos received more than 500 mm, whereas others only a few miles away appear to have received less than half that amount.

Methods

We made week-long field visits to the villages in September and December 1993; in April, September, and December 1994; in February, March, and September 1995; and in March, April, May, July, August, and October 1996. We also made several shorter visits throughout the period. We selected Mayo consultants from each village because they were known locally to possess superior knowledge of plants. None of the consultants was acquainted with those from other villages. All were paid at a rate higher than the prevailing day wage. We carefully explained the nature of the project, stressing that we considered the Mayos' knowledge to be of great importance, and asked them to teach us about their identifications and uses of plants.

The consultants accompanied us into the field, usually near their home village, but occasionally to more distant areas. We asked them to identify plants, stating both the Spanish and Mayo names, and to list and explain the uses of the plants. David Yetman and Rigoberto López Estudillo compiled extensive field notes independently and compared them later. Whenever possible, we verified the consultants' reports by asking identical questions of other consultants in another area. Vouchers of each ethnobotanical record were collected by Thomas Van Devender and deposited into the herbaria at the University of Arizona and the Universidad de Sonora. Plants were collected under a permit from the Secretaría de Agricultura y Recursos Hidráulicos in Navojoa. We identified specimens using comparative collections and literature in the University of Arizona Herbarium. We entered collection data, scientific and common names, and ethnobotanical information into a computer database.

Mayo Consultants

At Teachive, Vicente Tajia Yocupicio (born ca. 1931; fig. 4.3) dismounted from his bicycle to join a community plant-identification parade along the Arroyo Masiaca. Little by little, other villagers dropped out and Vicente remained, demonstrating his superior knowledge of plants, their anatomy, and their uses in healing. As we came to know him, he also exhibited an enthusiasm for plant identification and description of human uses that transcended the payment he received for his services. Vicente accompanied us on several field trips to remote areas, introducing us to new areas and to other consul-

FIGURE 4.3. Vicente Tajia Yocupicio of Teachive. Photograph by Thomas R. Van Devender.

tants as well. He enthusiastically taught us Mayo terms and spoke of Mayo customs, encouraging us to learn Mayo phrases as well. In discussing people of the region, he made a clear distinction between *yoremem* (Mayos) and *yoris* (mestizos).

Vicente was born in Teachive in a small house about 50 m from where he now lives. He earns his living at a variety of jobs, most frequently as a bagger of buffelgrass seed, backbreaking and choking work that is only intermittently available, for which he receives between $4 and $5 a day. On occasions he has refused to work in the distant fields of Huatabampo for 22 Mexican pesos (about $3 in U.S. currency as of this writing) because the cost of transportation to and from work would leave him with only a couple of dollars left over.

Vicente took advantage of our trips to acquaint himself with Mayos he had not previously known. At Las Rastras, Vicente was joined by Benjamín Valenzuela Moroyoqui (born in 1973), whose uncle, the late Luís Valenzuela

Moroyoqui (born ca. 1935; died 1997), was an expert on plant identification. Vicente had not previously met Benjamín but seemed quite delighted to make his acquaintance. On a later visit, Luís accompanied us up Arroyo Las Rastras and demonstrated a profound knowledge of Mayo terminology of plants. Because he was a bachelor, however, Luís had no family with whom he could share his knowledge of plant uses, and his memory of the applications of his field knowledge was limited. At Las Bocas, Vicente was joined by Paulino Buichilame Josaino, an acquaintance of many years.

Also providing valuable information at Teachive was the late Jesús José Moroyoqui (born ca. 1941; died 1997), who lived about 50 m from the house in which he was born. His wife, Lidia Zazueta (born ca. 1941), a rug weaver, was born in the house they share. Jesús worked sporadically in the community lands as a cowhand and sometimes worked as a field laborer in Mexican-owned agricultural fields near Huatabampo in the delta of the Río Mayo. He also shoveled gravel from the arroyo into dump trucks for roughly $5 a load. His family depended heavily on the income garnered from Lidia's weaving.

At Yocogigua, we retained the services of Santiago Valenzuela Yocupicio (born ca. 1916; fig. 4.4). He has lived in Yocogigua since he was a small child but was born at Las Rastras, approximately 16 km to the north. He lives with several children and many grandchildren in a house he built decades ago of native materials, including amapa, güiloche *(Diphysa occidentalis)*, mauto, palo colorado, pitahaya, and vara blanca.

Yocogigua is the site of an abandoned mescal factory, which flourished under a wealthy owner from the late 1920s until the early 1970s. The current building was completed in 1934 and produced a mescal bottled under the trademark Yocogigua. Mescal fanciers have commented to us on the product; their evaluations range from "horrible" to "undrinkable." In the 1970s the lands were confiscated and distributed as ejidos. For a decade the pueblo struggled to maintain production. According to Santiago, they ultimately agreed that the potent liquor was a bad influence on the people, and the enterprise was shut down. Now retired, Santiago worked at the factory for many years and as an agave gatherer, woodcutter, cowboy, and field-worker.

Braulio López (born ca. 1936) is president of the Yocogigua ejido, a job which requires him to work long hours as an administrator of the ejido's resources (fig. 4.5). He is widely respected and a prominent leader. He has worked with cattle most of his life. Braulio is recognized in Yocogigua as the individual who is most familiar with the plants in the area. In the field,

FIGURE 4.4. Santiago Valenzuela Yocupicio of Yocogigua. Photograph by Thomas R. Van Devender.

Santiago and Braulio would frequently have extended discussions, often in Mayo, about plant identifications and uses. In general, Braulio would defer respectfully to Santiago's age, whereas Santiago would defer respectfully to Braulio's sophistication.

Braulio's version of the closing of the *fábrica de mescal* is different from Santiago's story. According to Braulio, local residents started a vigorous bootleg operation within the factory, diverting considerable production to their own marketing system. Government investigators, he noted, smiling sheepishly, discovered the "clandestine" operation and shut down the factory, administering a serious blow to the town's economy.

At El Rincón, we had the distinct fortune to find Francisco Valenzuela Nolasco (fig. 4.6), known as don Pancho. (The term *don* is used as a mark of respect.) Don Pancho was born in 1933 or thereabouts in Los Tanques, some 25 km north of Alamos. (As is the case with many indigenous people, older Mayos tend to be vague when asked their age.) He has spent most of his life

FIGURE 4.5. Braulio López of Yocogigua. Photograph by Thomas R. Van Devender.

FIGURE 4.6. Francisco Valenzuela Nolasco of El Rincón, Sonora. Photograph by Thomas R. Van Devender.

in the monte or forest, primarily in the canyon and hill country adjoining his home. As a younger man he worked on the coast near Las Bocas on various construction projects. He now works several months of the year cutting poles of vara blanca. Among don Pancho's responsibilities as a member of Ejido Las Bebelamas are occasionally caring for cattle, maintaining fences,

FIGURE 4.7. *Left to right*, José, Blás, and Fausto, the López brothers of Nahuibampo. Photograph by Thomas R. Van Devender.

and reporting on supplies of vara blanca. Don Pancho has also built furniture from local plants, primarily guásima and batayaqui *(Montanoa rosei)*, and has carved ceremonial masks and spears from palo mulato and garambullo *(Pisonia capitata)*.

Don Pancho is isolated from other Mayos of the region. Although he still speaks fluent Mayo, it appears that his children and younger relatives do not. Although El Rincón was a Mayo settlement renowned for its fiesta only a few decades ago, the fiesta is no longer held, and Mayo customs appear to be dwindling.

At Nahuibampo, we retained the services of three brothers, José, Blas, and Fausto López (born between 1916 and 1935; fig. 4.7) and Genaro Liso Jusaino (born 1931). They were born in Nahuibampo and have lived there all their lives. They all work for Ejido Nahuibampo and are responsible for watching the ejido's cattle, a job of little responsibility in 1994 because most of the few cattle died or were sold due to the prolonged drought. Each has a milpa or *maguechi* (the Mayo name for a hillside), but the drought of the early 1990s made planting pointless. One of the brothers has become involved in intertribal discussions and hints that he has played an activitist's role as well.

FIGURE 4.8. Vicente Tajia Yocupicio's house in Teachive, Sonora (December 1993). The fence is made of split pitahaya *(Stenocereus thurberi)* ribs. Photograph by Thomas R. Van Devender.

Uses of Trees in Mayo Villages

Nahuibampo shares (with Yocogigua and Teachive) the features of most Mayo villages. Houses are set well apart on large clean-swept lots (fig. 4.8). Streets are demarcated by whitewashed stones placed at regular intervals. These stones are repainted at festival time, especially during Holy Week, when processionals bear testimony to the ancient Mayo culture. Homes have walls of plastered or unplastered adobe. Nonbearing walls, those not holding up heavy dirt roofs (fig. 4.9), may be of wattle-and-daub construction, typically of vara blanca, pitahaya, or batayaqui covered with mud plaster. Roofs are made of tough amapa and palo colorado poles covered with *latas*, closely placed slats of vara blanca or occasionally ribs from etcho or pitahaya, upon which some 25 cm of moist dirt is piled and smoothed. What roofs lack in elegance, they make up for in insulation, although they are also excellent habitats for scorpions and geckos. Most of the houses have grasses, shrubs, and even choya *(Opuntia fulgida)* or civiri *(O. thurberi)* growing on top of the roofs as testimony to the great age of the dwellings.

The inside walls of Mayo homes are plastered and whitewashed. Most floors are dirt; some are concrete; all are clean. Homes have wide porches with posts of amapa or mauto that are set on bases of palo blanco and support beams of amapa or palo colorado. Rooms are commonly separated by

FIGURE 4.9. Details of batayaqui *(Montana rosei)* walls. *A, above*, weathered adobe-plastered house wall, El Rincón, Sonora (April 1994). *B, opposite, top*, new unplastered house wall, El Rincón (September 1994). *C, opposite, bottom*, new unplastered wall of outside cooking area, Rancho Sierrita, Sierra de Alamos, Sonora (March 1994). Photographs by Thomas R. Van Devender.

a breezeway because of the intense heat of the long summer (May through October). Kitchens are open on one side, sheltered from wind and sun by a wall of interwoven batayaqui poles or split pitahaya ribs (fig. 4.10) that are covered with daub and sometimes plastered and are protected as well from rain and sun by a roof similar to that of the house. Both individual Mayo houses and Mayo villages are notable for their cleanliness, especially given (or perhaps because of) the ubiquity of roaming livestock: cattle, burros, horses, goats, sheep, pigs, chickens, turkeys, and ducks.

None of the homes that we saw had running water. In Teachive water is provided to a central outdoor spigot from a community-owned electric pump and well in the arroyo. Most homes in Teachive have electricity. In Yocogigua, the village and some homes are provided with a central spigot, and some houses have electricity. In Nahuibampo, some if not all of the water is hauled by bucket or *olla* (clay pot) from the Río Mayo, and only a few homes have electricity. At El Rincón, a gravity-fed piping system provides a

B

C

FIGURE 4.10. Detail of wall of outside cooking area of split pitahaya *(Stenocereus thurberi)* ribs, Huebampo, Sonora (September 1994). Photograph by Thomas R. Van Devender.

central faucet, and each house has modest plumbing but no electricity. Each home in each village has a large ceramic olla filled with water for drinking. It rests on a *tinajera* or *horqueta* (tripod or forked pole) made of the trunk and stout branches of mauto, mezquite, or other hardwood (figs. 4.11 and 4.12).

Results and Discussion

The investigation was predicated on the assumption that insights gained from native peoples whose lives are intimately tied to the land can provide important scientific knowledge of and innovative ways of looking at plants. Native peoples' uses of trees and tree materials penetrate to the heart of human material culture, illuminating its ethos and ultimately the soul of its people (see, for example, Felger and Moser 1985).

Trees are quite diverse in Sinaloan thornscrub and TDF and are important natural resources for the Mayo. Ethnobotanical uses—even if nothing more than fence posts or fodder for livestock—were noted for 92 species of trees in 35 families (appendix 4.1). Only palo joso de la sierra, the tallest tree in the TDF at 20 m or more, was emphatically said to have no use at all. All those familiar with it admired its stature, flowers, and dignity but had little more

to say. Even so, we have included it in appendices 4.1 and 4.2 because its great beauty seems to be an important use. The uses of trees by the Mayos fall into the following broad categories.

1. Esthetic uses: trees used in their unmodified conditions. Cascalosúchil or candelilla *(Plumeria rubra)* is valued as an ornamental. Guamúchil *(Pithecellobium dulce)* is valued and praised, not merely for the fruit it bears, but also because it is an excellent tree. Palo jito *(Forchhammeria watsonii)* is valued for its shade. Gloria *(Tecoma stans)* appears as an ornamental in nearly every yard. Its leaves and flowers are also valued for their medicinal properties.

2. Livestock feed uses: trees used for their ability to feed and nourish livestock. Many legumes (mezquite, brea, palo fierro, and palo zorrillo) produce edible pods and branches. The fallen corollas of some trees (guayacán and palo santo) are consumed by livestock. The bark and inner wood of palo santo and güinolo are eaten by burros in times of drought. The foliage of many legumes nourishes goats. Several species (chapote, mezquite, and palo zorrillo) provide pollen and nectar for honeybees. Goats

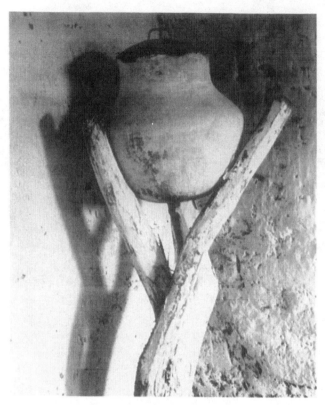

FIGURE 4.12. Water olla supported in horqueta of mezquite *(Prosopis glandulosa)* wood, Yocogigua, Sonora (December 1994). Photograph by Thomas R. Van Devender.

generally eat all species consumed by cattle, plus several others. All livestock uses are post-Conquest innovations.

3. Livestock management uses: trees used inherently or as materials for managing livestock. Short trees with bare boles (guayacán) provide roosts for chickens, where they are safe from predators, including dogs. Mayos provide ladders for the chickens to make access to the roosts easier. The woods of other trees are valued for fence posts (fig. 4.13): some (chilicote, etcho, jaboncillo, and torotes) for providing living fences; others (brasil, mezquite, pitahaya, and tepeguaje) for building corrals or fencing around homes.

4. Construction uses: trees used in building. Certain hardwoods (amapa, mauto, palo blanco, and palo colorado) are especially valued for their durability and resistance to damage by water and insects. These qualities make them desirable as roof beams *(vigas)* and posts *(horcones)*. Although

FIGURE 4.13. Fence of pitahaya *(Stenocereus thurberi)* ribs, Masiaca, Sonora (October 1994). Photograph by Thomas R. Van Devender.

less durable, other woods are easily cut and are resilient (batayaqui, pitahaya, and vara blanca), making them useful for cross-hatching in roofs and for wattle (figs. 4.9 and 4.10).

5. Artifact uses: trees and tree parts used in making tools, implements, furniture, utensils, etc. Palo de asta is especially noted for its durability in tool handles. Guásima is the chosen wood for chairs and tables (see fig. 1.2). Guayacán is used in building looms. Mauto is used to make stands for water jugs (tinajeras; fig. 4.11). Palo chino *(Havardia mexicana)* is carved into bowls and ollas, as are the burseras. Sabino is made into fine furniture and cribs (fig. 4.14). Alamo *(Populus mexicana)* is cut into thin, wide boards for table tops.

6. Industrial uses, i.e., tree products consumed in producing other materials. Dyes are made from brasil and palo dulce *(Eysenhardtia orthocarpa)*. The bark of mauto, palo chino, and palo joso is used to tan leather. Samo baboso is added to lime to make it adhere to plaster or adobe. Tempisque bark or sap is used to coagulate milk in making cheese. Numerous trees are used as firewood. The selection is typically dependent on the temperature of fire desired. The roots of pitahaya were formerly burned to produce ashes used in making soap.

7. Medicinal uses: plant parts used to combat illness or maintain health. All

FIGURE 4.14. One-year-old Manuel Esteban Campoy Valenzuela in 100-year-old crib made of sabino *(Taxodium distichum)* wood, Yocogigua, Sonora (September 1993). Photograph by Thomas R. Van Devender.

parts of the plant are used (root, wood, bark, branches, leaves, flowers, fruits, and seeds). The sap of torote copal is chewed as a gum; that of brea is used to relieve bronchitis and asthma. The root bark and sap of tescalama *(Ficus petiolaris)* are used to cure hernias. Some products are applied directly to wounds or injuries (juice from etcho). Others are prepared as a tea or eaten raw, cooked, or in combination with other materials, both plant and animal. Some plant parts are sliced; others are mashed, twisted, or ground. Some are heated; others are toasted, roasted, or used at ambient temperature. Some are steeped in alcohol; the fumes are inhaled or the liquid is drunk. The medicinal uses of plants include veterinary uses as well because livestock are economically critical for many Mayos.

Some products are deemed effective remedies for human conditions that are not recognized by Western science. These infirmities include *susto* (a fear or anxiety), *empacho* (digestive distress related to imbalance in a physical system in the body), *pasmo* (an apparent neurasthenia connected with physical or emotional trauma), *tirisia* (sadness or melancholy), and conditions variously referred to as "hot" or "cold" and usually related to reproductive dysfunction. Some products are credited with antibiotic properties (copalquín and palo mulato).

The ubiquity of remedies for intestinal disorders indicates a high incidence of parasitism. The wide variety of remedies for insect bites and stings indicates an environment rich in poisonous invertebrates, a fact attested to by the researchers. The remedies suggest several natural insecticides, miticides, antihelminthids, and antiprotozoans.

8. Food uses: plant parts consumed as food as opposed to medicinal or cultural consumption. Many fruits are eaten raw, others are cooked, and some require elaborate preparation. The root of pochote is eaten as a starch in times of drought. The barks of some trees (palo mulato and torote copal) are brewed for refreshing or stimulating teas. The seeds of giant cacti (etcho and pitahaya) are valued as a source of starch. An *atole* (beverage) is made from etcho seeds. Many legume fruits are regarded as edible, although of questionable virtue.

9. Cultural and religious uses: plant parts incorporated into Mayo culture because of their cultural or spiritual significance. Although most of these uses are derived from shrubs and herbaceous plants, some trees are also used for these purposes. Abolillo *(Sapindus saponaria)*, and palo dulce are used to make beads for rosaries, necklaces, and crucifixes. The wood of brasil is chosen to make points for the lances carried by *fariseos* (Pharisees) in Lenten processions. Sitavaro *(Vallesia glabra)* is used to counter spells or infections from the dead. It is taken as a tea to ward off invasion of ethereal pathogens through boils and sores. Alamo branches are used to deck ramadas and procession routes during Holy Week. Specific herbs may be prescribed by a *curandero* (healer) to combat a spell placed on one person by another.

This list and the specific uses reported in appendix 4.1 reflect the pervasiveness of native plants, especially trees, in Mayo culture. Most of our consultants were older men. Older women also appear to be highly sophisticated in plant identification, but because of division of labor, they spend a disproportionate amount of time in the home and have far less opportunity to become familiar with plants in their native habitat. With the exception of firewood, which everyone collects, it is usually the men who collect the desired plants in the monte and bring them to the houses. There, construction and wood carving are done by the men, whereas the women process the plants used for food, medicine, dyes, etc.

We have encountered few younger Mayos with sophisticated field knowledge or intimate applied knowledge of plants and their uses. Benjamín Valenzuela Moroyoqui of Las Rastras, born in 1973, is the only younger person we found willing to guide us into the monte to describe the plants. This does not imply that other such young people do not exist, for our experience has taught us that many youths living in rural Sonora carry with them immense knowledge of plants and their uses. It does indicate, however, that there are fewer younger people involved in the study of plants and their traditional Mayo uses. Most villages support few young men; they are forced by economic necessity to find work elsewhere. Once they become assimilated into city life, they lose their close identification with the land and its products. Traditional Mayo knowledge of plants is not being passed on to younger people in many places. Once the children begin school at the age of five, they are expected to speak Spanish instead of Mayo. More subtle uses of plants for food and medicinal remedies may be lost in a generation or so, replaced by food from the *tienda* (store) and medicines from the *farmacia* (pharmacy). In contrast, plant uses for construction, livestock, etc., that are shared with the Mexican mestizo culture will persist longer than the Mayo language. There is a great need to assure the Mayos that their knowledge of plants is valuable and should be passed on to future generations.

Our investigation reveals many towns on the verge of economic calamity. For reasons beyond the scope of this paper, the only local employment available to men consists of woodcutting and caring for the few cattle the pastures are able to sustain. The villages demonstrating the worst economic conditions tend to be those in which the citizens profess little or no knowledge of Mayo language and culture. Many deny any connection to Mayos, although their physical characteristics and their geographical histories virtually guarantee their Mayo ancestry. In one case, a young mother vehemently denied any knowledge of plants or the Mayo language but conversed fluently in Mayo with her mother when the latter responded to our questions.

In villages with Mayo history but scanty or nonexistent Mayo presence, we hired men reputed to be the most knowledgeable about plants to guide us into the monte. Although they were congenial fellows, their knowledge of plants was less comprehensive than that of individuals in towns where the inhabitants were more clearly willing to state that they were yoreme.

In the background of the region of our investigation lies the ubiquitous drug trade. The agronomy and marketing of *mota* (marijuana) are critical to the economy of the region, although for obvious reasons statistics on the extent of the traffic are unavailable. Suffice it to say that local reports assert the pervasiveness and economic importance of planting, raising, harvesting, transporting, and merchandising drugs. As one Mayo pointed out, "When there is no work, the young men decide to go into crime, mota, that sort of thing. That's what's available."

ACKNOWLEDGMENTS

Victor W. Steinmann, Richard S. Felger, and Philip D. Jenkins helped identify ethnobotanical specimens at the University of Arizona. Carl A. Olson, curator of the University of
Arizona entomology collection, provided names for the Mexican jumping bean moth and
the goma de Sonora lac insect. Financial support for this project was provided by Native
and Nature, the University of Arizona Social and Behavioral Sciences Research Institute,
the University of Sonora–University of Arizona Collaborative Grants, the Jacobs Fund, and
the Roy Chapman Andrews Fund of the Arizona-Sonora Desert Museum. Anne E. Gondor
drafted the map. Thomas R. Van Devender took all of the photographs.

APPENDIX 4.1
Ethnobotanically Useful Trees and Columnar Cacti of the Mayo Region

---

(M) = Mayo common name; other common names are English or Spanish.

APOCYNACEAE

Plumeria rubra L. f. Candelilla, cascalosúchil. A small, irregularly shaped tree with blunt
branches that explode into exquisite white blossoms in late May, June, and July. The brilliant
white of the blossoms contrasts handsomely with the green of the leafage. Mayos recognize
the tree for its beauty and are said to have planted it near their houses.

Vallesia glabra (Cav.) Link. Sitavaro (M). Commonly a shrub or vine bearing tiny whitish
fruits the size of peas, it grows into a quickly spreading tree nearly 6 m tall in the foothills
of the Sierra de Alamos.

The milky sap is highly recommended for cloudy or infected eyes. The sap exudes from
broken fruits or branches, is carefully applied to the corner of the eye, and is said to cure infections rapidly. The root is boiled to produce a yellow-brown dye. The plant is considered
to be a powerful anti-infective in the sense that some physical pathology is believed to be
communicated by the dead. Tea from sitavaro branches is used to wash sores or boils, not
only to help them heal, but also to prevent them from contracting agents of pathology from
the dead.

BIGNONIACEAE

Tabebuia chrysantha (Jacq.) Nichols. Amapa amarilla, to'obo saguali (M). A strong-boled
tree similar in configuration and leaf structure to the amapa. However, because it has been
persecuted for its wood or because it is of smaller habit, amapa amarilla appears smaller
than its sister. It is nowhere common. It is limited to scattered individuals on slopes in the
TDF, usually distant from human habitation, suggesting a history of heavy human exploitation. Its absence from villages is unfortunate indeed, for in March and April the tree bursts

into cascades of yellow, whose brilliance is unmatched in the world of flowers. To stand beneath an amapa amarilla in full flower is to be transported to an unworldly realm of splendor.

Mayos attribute the same qualities to its wood as to that of the amapa (see below), except that it is reputed to be even more durable. This is difficult to verify, however, because it seems doubtful that the tree ever grew in sufficient numbers to constitute a significant source of lumber. Still, all of the consultants expressed an admiration for the tree and identified locations where it grows. When not in flower, it can be distinguished from the pink amapa by its lighter bark and rougher leaves.

Tabebuia impetiginosa (Mart.) Standl. Amapa, to'obo (M). A strong spreading tree with dark, rough bark, up to 12 m tall in TDF. In winter the bare branches explode into a riot of pink blooms, enhancing the forest in all directions. The profuse and colorful blossoms, a blessing to humans, have been a curse to the species, marking the trees as easy targets for woodcutters and pushing the species to near extinction in some areas. Untold millions of these trees were cut for roof beams for buildings and for mine timbers that propped up hundreds of miles of tunnels from which the silver of Alamos was extracted. The Mexican government has wisely chosen to protect the tree. Mid-sized individuals are once again common in the forest, where the slopes are dotted in winter with pink accents.

The stout, heavy bole of dense wood is the best stock for posts and beams for homes, resisting rot and infestation by boring insects. Most older homes in its habitat feature *vigas* (beams) and *horcones* (forked posts that support vigas) of amapa. Beams from abandoned houses, often more than a hundred years old, are eagerly collected for use in new buildings. Ashes of the wood can be used like lime to soften corn for grinding into flour.

Tecoma stans (L.) Juss. ex H.B.K. Gloria. A small tree with numerous brilliant yellow tubular flowers that contrast most agreeably with its dense green foliage. It is uncommon in its wild state in the Mayo region; we found only one wild specimen during the fieldwork. The leaves are widely believed to have remarkable medicinal powers, especially for lowering blood pressure.

BIXACEAE (COCHLOSPERMACEAE)

Cochlospermum vitifolium (Willd.) Spreng. Buttercup tree, palo barriga (widely called palo barril elsewhere), ciánori (M). A most imposing tree with a thick trunk covered with silvery gray bark that resembles elephant skin. The smooth, symmetrical bole rises to considerable heights, typically in excess of 15 m. In late winter when it is leafless, the tree gives forth brilliant yellow, cup-shaped flowers up to 10 cm in diameter, which explode from the tips of the branches. Seen from a distance, they resemble golden yellow ornaments. In late fall the broad leaves turn to effusive shades of red, yellow, and gold that are reminiscent of more northern autumns.

The wood is carved into a basin and filled with water, which is allowed to stand overnight. Women drink the water to alleviate varicose veins.

BOMBACACEAE

Ceiba acuminata (S. Watson) Rose. Kapok, pochote, baogua (M). The kapok tree grows to more than 15 m. Its branches often develop in linear zigzags, and its large, palmately compound leaves illuminate the landscape with a light green that turns yellow in early fall. Individual trees become covered with large imposing thorns in youth. They are innocuous, however, becoming pyramidal, softening with age, and lending an air of the exotic to the trees. The fruits usually develop in spring and burst open in fall in the form of softball-size puffs of cotton, which are conspicuous from great distances on the leafless trees.

A vessel of the wood is filled with water, and the liquid is drunk as a cure for varicose veins. The puffy fibers are used as stuffing in pillows and as cushioning in cribs. In times of scarcity the roots of the tree are eaten with great enthusiasm by native people. The wood is used for making cots. Recently the women of La Aduana near Alamos have begun using the bark as stock for carving miniature landscapes of the type introduced many decades ago in southern Mexico.

Pseudobombax palmeri (S. Watson) Dugand. Cuajilote. A gnarly, contorted tree seldom more than 7 m tall, with ruby wine-colored bark that peels into concave sections of an attractive mottled red color. It flowers when leafless in April, presenting pollinators with large, sensuous, upright blooms with a multitude of stamens that spill lasciviously over the edges of the brilliant white blossoms. The flowers open at night and are visited immediately by nectar-eating bats and moths. The large leaves turn gold in the fall, retaining their vigor and presenting a sharp outline of color until they fall to the ground, still strong and entire.

The wood is carved into a bowl into which water is poured and drunk. According to some Mayos, it will cure varicose veins. The tender fruits are also eaten. Our Mayo consultants were familiar with cuajilote, even though they lived outside of its range.

BORAGINACEAE

Cordia sonorae Rose. Palo de asta, pómajo (M). A slender clean-barked tree up to 10 m tall with a smooth, silvery-gray trunk. It leafs and flowers toward the tips of its branches. It flowers a delicate white in late winter; the individual blossoms are visible from afar as white points in the forest. The Spanish name means "flagpole," suggesting the rapid, straight growth of the young trees.

The hard, vigorous wood is widely used for posts, beams, and vigas in houses and is the main choice for handles for tools and implements. Few Mayo homes lack an implement with a pómajo handle.

BURSERACEAE

Bursera fagaroides (H.B.K.) Engl. var. *elongata* McVaugh & Rzed. Torote, torote de venado, torote de vaca, to'oro (M), to'oro saguali (M). A prominent tree during the leafless season, growing to 8 m tall. The light-colored bark ranges from sandy to light green and is usually colored with layers of wispy exfoliations that whistle in the wind. The wood is used for fence posts and living fences. The wood is also carved into dishes and utensils, but according to some Mayos, is inferior for this purpose.

Bursera grandifolia (Schlect.) Engl. Palo mulato, to'oro mulato (M). Palo mulato is a handsome spreading tree up to 10 m tall that is much esteemed by natives. Its bark is dark bluish green to silvery gray to red, with tan exfoliations that give it a distinct appearance.

The thick reddish inner bark is relished as a tea and as a coffee substitute. In Nahuibampo, it was the only beverage available to the Mayos, who had no money to purchase cof-

FIGURE 4.15. Mayo mask carved from torote prieto *(Bursera laxiflora)* by Benigno Buitimea Acosta of Teachive, Sonora, in 1993. Drawn by Anne E. Gondor from the original.

fee. Tea from the bark of palo mulato is widely used as a tonic that is said to restore energy by thickening the blood. Locals claim that the redder the bark, the more effective the remedy. Certain individual trees have heavily scarred trunks, indicating they are known to have especially potent bark. The white wood is also used to carve masks, utensils, and bowls. The meaning of the Mayo *mulato* is obscure. One source asserted that it means "tea maker," but others reported that the name is merely very old with no obvious meaning.

Bursera lancifolia (Schlect.) Engl. Torote copal, to'oro chutama (M). A large spreading tree up to 10 m tall with small lance-shaped leaves. Its dark gray, exfoliating bark exudes a hint of incense when freshly cut. The sap is chewed for toothache and is used as a resin for the bow of a violin. The fragrant bark, steeped in hot water and drunk as a tea, is used to cure cough. Smoke from the wood is said to cure headaches.

Bursera laxiflora S. Watson. Torote prieto, to'oro chucuri (M). A large spreading tree with smooth, dark gray bark and tiny leaves. The ends of the branches are elongated and bear a reddish hue. The bole often becomes distended with age. The bark is used to cure cough. The wood is carved into masks, spoons, and trays (fig. 4.15). At Teachive branches of the tree are placed at the bottom of a fire pit when agaves are roasted. The aromatic wood is said to impart a pleasant aroma while the agave is cooking and add a sweetness to its flavor.

Bursera penicillata (Sessé & Moc. ex DC.) Engl. Torote acensio (= de incienso), to'oro (M). The tree exudes a fragrance detectable from many yards away. The leaves give forth an odor like anise. The investigators have often had the experience of noting the fragrant smell of the tree before seeing it. The blunt ends of its branches help distinguish this species from others. Torote acensio commonly grows to great size (some individuals exceed 15 m) and is said to be the largest of the region's common burseras. The wood is excellent for carving into bowls and utensils.

Bursera stenophylla Sprague & Riley. Torote copal, to'oro (M), to'oro sajo (M). The small, dainty leaflets help distinguish this species from others. The sap is chewed as a gum and is said to fight off colds. The wood is carved into bowls and utensils. Individual trees grow to large size; some reach 75 cm in diameter.

CACTACEAE

Carnegiea gigantea (Engelm. ex Emory) Britton & Rose. Saguaro, saguo (M). The southernmost population of this Sonoran Desert giant cactus is found in large numbers on the basaltic sides of Mesa Masiaca, only a few kilometers from Teachive. Despite the nearness of the population and the importance of the saguaro in the lives of other indigenous peoples, our Mayo consultants stated they knew of no important purpose for the plant and expressed no interest in it. They find the fruit to be inferior to the fruits of other local cacti. The ribs, used widely by native peoples to the north, are weaker and shorter lived than those of etcho or pitahaya.

Pachycereus pecten-aboriginum (Engelm.) Britton & Rose. Etcho (M). The Mayo name *etcho* has often been confused with the Spanish word *hecho* (fact, made). This stellar tree is greatly esteemed by Mayos, who named one of their major settlements, Etchojoa, after it. Specimens exceed 10 m in height and sport numerous arms, becoming truly giant cacti. The fruits, although they appear to be covered with sharp spines, are in fact relatively harmless and are said to have been used by Indian women to comb their hair; hence, the scientific species name. In the searing drought of late spring, etchos are often the only trees on the hillsides retaining a vestige of green.

The strong ribs of the cactus are used in house construction and in making looms and other household artifacts. The wood is carefully sawed and shaped into boards for benches, beds, and roof beams. Juice from the flesh is squeezed from a section and drunk for *mal de orin* (bad urine) and for *la prostatitis* (prostate trouble). To stanch bleeding, a few drops of the juice are dripped onto the wound and a *gajo* — a piece from a cross section — is placed directly over the wound.

The fruits of etcho are eaten raw or cooked and are made into wine and jelly; they are also dried and preserved. The raw seeds were formerly ground into a flour. Dissolved in water, they were drunk as an *atole* (beverage). Mixed with a small amount of corn flour, they were said to make excellent tortillas. Older Mayos reported that they regularly ate etcho seed tortillas as children. Some women still prepare tortillas and atole from the seeds, as doña María Teresa Moroyoqui (wife of Vicente Tajia Yocupicio of Teachive) did for us, but the practice is vanishing. Small etchos are planted as fence rows and grow quickly (within 10 years) into an impenetrable barrier, a practice that long predates the introduction of barbed wire. Its strong trunk and spreading habit also make etcho valuable as a shade tree, an uncommon virtue for a cactus.

Stenocereus montanus (Britton & Rose) F. Buxb. Sahuira (M). A stately, uncommon, 10-m-tall columnar cactus typically found in higher, well-developed TDF. It usually grows several large arms and superficially resembles the etcho. In northern Sinaloa it becomes the dominant arboreal cactus. The fruits are eagerly eaten and are said to be superior to the delectable fruits of the pitahaya. The wood is said to be more durable than that of the etcho. Because of its small numbers in Mayo country, however, its use is limited. Some Mayos expressed a desire to cultivate the plant.

Stenocereus thurberi (Engelm.) Buxb. Organpipe cactus, pitahaya, aaqui (M). A common aborescent cactus of Sonora that grows to great size in the TDF (as tall as 10 m), often with hundreds of arms. Larger individuals possess a thick trunk that rises a couple of meters above the ground. The tree grows in enormous numbers and in dense forests on the coastal plain south and east of Huatabampo. The individuals there, however, do not attain the great stature of those in more mesic habitats.

The ribs are commonly used for fences, walls, ceilings, and some furniture. Formerly, large numbers of cuttings were planted in rows to produce living fences. The lower wood of the trunk and arms is surprisingly sturdy and is used in many phases of building, including cross-beams bearing heavy weights. The delicious fruits are relished and eaten in immense numbers in August and September, constituting an important component of the Mayo diet and a potentially valuable economic resource. Formerly they were dried and preserved or made into wine. A woman from Masiaca reportedly makes tamales from the fruits. The fleshy, moist stem is singed to remove spines, then applied directly to the flesh for snake and insect bites, a remedy that was tested with positive results when Tom Van Devender was bitten by an assassin beetle. The burned peel of the fruit is said to be applied to the anus for hemorrhoids. Dried peels are boiled into a tea and taken by women to stop hemorrhaging.

CAPPARACEAE

Forchhammeria watsonii Rose. Palo jito, jito (M). A gnarled, strong-trunked, symmetrical tree up to 9 m tall (sometimes taller), often resembling a lollipop in shape. It is endemic to southern Sonora and northern Sinaloa. It retains its dark green cover throughout the year, leafing out anew in late spring so that even in the desiccating blasts of June its leaves are a dark, shiny green. Gentry (1942) says "The tree has a very individual appearance, suggestive of old olive trees in ancient Judea. In the burning days of late spring it is about the only tree that offers shade to weary beasts and man." Individual trees are respected and recognized. One tree near Teachive that is more than 10 m tall with a trunk of more than 1.5 m in circumference is known as "El Jitón." It is widely renowned, and the shady ground beneath it is used as a resting place for foot travelers. Each year large jitos yield many kilograms of fruits, which when tender, are boiled with sugar and eaten with great relish.

CELASTRACEAE

Wimmeria mexicana (DC.) Lundell. Algodoncillo, chi'ini (M). A small tree in much of the forest but grows to 10 m tall in some areas, often with multiple trunks from a common root.

The trunk is thick muscled and strong. With age, the bark peels away from the trunk in small sections, sometimes leaving a multicolored mosaic pattern on the exposed bare trunk. When a piece of this deciduating bark is carefully broken, tiny cottony fibers appear, holding the pieces together; hence, the Spanish name, which is derived from *algodón* (cotton). The hard trunks are used for fence posts and for beams in homes. They must be replaced more frequently than other woods such as chopo because they are said to be more susceptible to rot and infestation by termites.

COMPOSITAE (ASTERACEAE)

Montanoa rosei B.L. Rob. & Greenm. Batayaqui (M). A slender, flexible shrub or tree about 5 m tall with thin, straight, and smooth stems, growing in extensive, often impenetrable thickets. In January the trees produce a profusion of dense white blooms that permeate the region with a most agreeable perfume. Batayaqui grows quickly and is in abundant supply despite heavy use by natives.

One of only two tree composites known from the area, batayaqui is a most versatile, widely used tree. *Tarimas* (beds) called *tapestes* (Cáhita) are made by lashing lengths of poles together. Such a bed was demonstrated at El Rincón and proved remarkably comfortable. The wood is widely used to make backs for chairs, animal cages, and crates for shipping. The springy poles are woven to produce a sturdy wall or roof, which may be plastered or not. Arrows are still made from the straight stalks. The gum is applied to sores as a putative healing agent.

CONVOLVULACEAE

Ipomoea arborescens (Humb. & Bonpl.) G. Don. Morning glory tree, palo santo, jútuguo (M). A broad-boled tree up to 12 m tall that quickly tapers into narrowing branches, which decurve gracefully and randomly from the gray-white trunk. (At higher elevations the trunks of the variety *pachylutea* are yellow.) During the rainy season, the large ovate leaves provide dense shade. During the dry season, the tips of the bare branches burst into large white flowers that stand out from afar.

The fallen flowers are eagerly consumed by livestock and deer. They are also reportedly effective at controlling high blood pressure when brewed into a tea. The bark and wood are eaten by livestock, especially burros, which in years of drought (i.e., most years) causes heavy attrition in plant numbers. The bark from branches is steeped into a tea that is said to be a dramatic remedy for insect bites, offering relief within an hour of the drinking. If a drop of the congealed sap is placed on an aching tooth, the pain is said to cease almost immediately. Natives warn, however, that care must be taken not to touch healthy teeth because the affected tooth will slowly disintegrate over the following few weeks. Two of the consultants indicated bare gums where they had treated diseased molars with the sap of jútuguo. Large strips of the bark are peeled from the tree and used as thatch for roofs.

EBENACEAE

Diospyros sonorae Standl. Sonoran persimmon, guayparín, caguorara (M). A straight-boled, uncommon tree, reaching 15 m tall with a spreading habit. It is valued for its fruits, which are eaten cooked. Many bird species flock to the trees to consume the ripe fruits. Locations of the trees are well known among Mayos, who eat the fruits and grind the seeds to produce atole.

Erythroxylaceae

Erythroxylon mexicanum H.B.K. Mamoa (M). Normally a many-trunked shrub, in rich TDF it grows into a tree up to 8 m tall. The roundish leaves, which ordinarily do not drop from the tree in the dry season, harbor a hint of iridescence, helping to distinguish the tree from *Karwinskia humboldtiana*, with which it is easily confused.

The wood is said to be strong and durable, good for making fence posts and roof supports. Smaller trunks are used as *latas* (crosspieces) in roofs to support the layer of earth above.

Euphorbiaceae

Croton fantzianus Seymour. Vara blanca, cuta tósari (M). A common small tree up to 8 m tall that once grew in thick stands. Populations of mature trees have been decimated by cutting. The leaves turn yellow, orange, and red in senescence, presenting an autumnal lilt to the landscape well into the winter months.

The trees are cut into stakes and used in many aspects of house building, especially as latas for roofs before dirt is applied. Lengths slightly longer than 2 m are sold by the millions as stakes for tomato and grape crops in Baja California. Cutting the vara blanca provides many jobs in the TDF of Sonora and Sinaloa.

Croton flavescens Greenm. Vara prieta, júsairo (M). A small tree, reaching little more than 3 m in height. In fall its leaves turn a mottled red, adding an attractive color to the thornscrub and TDF. The bark brewed into a tea is widely used to treat stomach ailments.

Jatropha cordata (Ortega) Müll. Arg. Torote papelío, panalero, to'oro (M). A striking, irregularly branched, narrow tree reaching 8 m in height, with soft wood and a green to yellow trunk, from which peel large sheets of exfoliating tan bark. In July the dark green leaves are accentuated by delicate pink flowers the size of lilies of the valley.

According to the Mayos of Teachive, large rectangular sheets of bark were carefully cut from the live tree and used to wrap *panal* (fresh honey) from wild bees for shipping and storage. The bark is still said to preserve freshness in the honey and comb. A tea is made from the bark, and the liquid is rubbed on sore kidneys. The tea is also applied to bee stings, instantly alleviating the discomfort, or so it is said.

Sebastiania pavoniana (Müll. Arg) Müll. Arg. Mexican jumping bean, brincador, túbucti (M). A small tree seldom more than 8 m tall. Its smooth, elongated shiny leaves of dark green turn red in the late fall and winter, making it easy to spot on the hillsides from ground or air.

The triangular seeds are used as toys. The movements of a moth larva (*Cydia deshaisiana* (Lucas), Tortricidae) cause the seeds to tumble and turn of their own accord. Formerly found in curio shops on the U.S. border, Mexican jumping beans are harvested sporadically and sold to vendors.

Fagaceae

Quercus chihuahuensis. Encino. A spreading oak tree up to 10 m tall but commonly smaller. Normally a tree of the higher slopes, it was found at about a 500-m elevation in El Cajón del

Sabino, where it was identified by don Pancho. Because it seldom grows in the TDF, he was unfamiliar with its uses except as firewood and the acorns as food for *güíjolos* (wild turkeys). In that area, however, pigs eagerly gobble up the acorns.

FOUQUIERIACEAE

Fouquieria macdougalii Nash. Tree ocotillo, jaboncillo, palo pitillo, murue (M). Shrubby in the desert, the plant becomes a tree in the Mayo region, sometimes rising as high as 10 m in a straight trunk covered with exfoliations and a dark mosaic. The spine-covered, greenish, exfoliating bark and the bright red flowers are identifying characteristics.

The fast-growing trees are planted to make living fences. The bark is widely used as a source of shampoo. Slices of the bark applied to a cut will supposedly stanch the bleeding and promote healing. One seldom sees large specimens of this large plant that have not been harvested in one way or another.

LEGUMINOSAE (FABACEAE)

Acacia cochliacantha Humb. & Bonpl. ex Willd. Boatthorn acacia, güinolo, chírahui, chírajo (M). An extremely common tree seldom exceeding 8 m in height. In parts of the forest, it thrives in uniform stands as second growth. It is an important plant in the succession of disturbed soils and seldom lives in these successions for more than a couple of decades. The thorns turn gray and hollow with age. As Gentry (1942) noted, in death the trees assume a roughly circular shape, the branches curling over upon each other as though the dead tree were rolling into a coil. When numbers of these skeletons are seen together, the landscape takes on an eerie tableau of death. It is one of the few plants able to compete successfully with buffelgrass in fields cleared of TDF.

The spines or thorns are brewed into a tea that is highly recommended for prostate and urinary problems. The legs of benches at San Antonio, north of Teachive, were fashioned from the tree. The wood is used for firewood. In times of scarcity, burros strip long sections of bark from the tree and apparently consume it to their benefit. Most larger trees near homes bear heavy scars.

Acacia coulteri Benth. Guayavillo, baigüío (M). In parts of the TDF, this tree resembles *Lysiloma divaricatum* in habit, sometimes growing as tall as 10 m. An exceptionally strong-trunked tree, its wood is ideal for beams and posts. It is often found in groves but is absent from large portions of the forest, a fact perhaps explained by its popularity as a lumber wood. It can tolerate more arid conditions than mauto, growing abundantly on Mesa Masiaca, one of the most xeric environments in the region.

Acacia farnesiana (L.) Willd. Sweet acacia, vinorama, cu'uca (M). Normally a large spreading shrub, cu'uca grows into a stately 10-m-tall tree in the Río Mayo region. The wood is used to create a blue dye, which is produced by prolonged soaking of the wood in water. The root is cooked as an antivenin for invertebrate stings and snakebites. In late winter the tree produces a profusion of round yellow blossoms that attract humans (as well as bees) with their perfume. Cu'uca also produces a tolerable firewood.

Acacia occidentalis Rose. Teso (M). A large, solitary, spreading acacia reaching 10 m in height and flowering white in late winter. It grows in pockets on low-lying flats, where it is quite common, but is absent in other areas that are ecologically similar. The hard wood is quite good for ceiling beams (when straight lengths can be found) and posts and is carved into *bateas* (wooden bowls).

Albizia sinaloensis Britton & Rose. Palo joso, joso (M). A graceful legume with a straight, yellow-white bole rising to 15 m in height, where it branches into a canopy of lacy greenery. It is valued for its beauty in inhabited areas. The wood is used for making utensils, such as spoons and bowls. At San Antonio the seat of a bench had been shaped from joso. At Masiaca the bark is soaked and used to produce a white-colored leather, which is marketed commercially.

Brongniartia alamosana Rydb. Palo piojo, vara prieta (near San Bernardo), cuta nagüía (M). A very common small tree in the TDF. The slender trunk rises irregularly to 8 m, and the bark is pocked with tiny lenticels, which give rise to its name, the "louse tree." It flowers dark blood-red in July during the height of the dry season. Its drying fruits explosively dehisce in the fall, often causing a temporary excitation of the human nervous system. *Nagüía* in Mayo means "good for nothing." It is also Mayo slang for a homosexual. Nevertheless, the wood makes acceptable firewood.

Caesalpinia caladenia Standl. Palo piojo blanco, jícamuchi (M). An elegant 6-m-tall tree, straight-boled, that is absent from some areas but remarkably common in others. The thin, light gray bark is underlain by bright green, which is revealed when the bark is scratched, a means of identification used by the Mayos. Flowers are pale yellow; beans are black. The wood makes acceptable firewood. It was also reported to make acceptable vigas if cut during the full moon.

Caesalpinia platyloba S. Watson. Palo colorado, mapáo (M), cuta síquiri (M). A strong-trunked, dense-wooded legume with large clusters of persistent broad pods that turn reddish brown with age. The mottled bark typically retains a reddish tint. The leaves are also somewhat more persistent than those of other legumes and assume a reddish color in senescence. The wood is acclaimed as the best for fence posts, reputed to last 100 years. It is also used in vigas because it resists rot and insect infestations. It is a most esteemed and valued tree in Mayo country. Its numbers have been decimated owing to its excellence in home construction and fence building.

Caesalpinia sclerocarpa Standl. Ebano (M). A spreading, 12-m-tall tree with very dense wood and smooth mottled bark. Its wood is said to be the very best available in thornscrub for construction of houses. Perhaps because of overharvesting or the natural limits of its distribution, ébano grows in Sonora only in Arroyo Camahuiroa, where four large trees are found. Natives say it is more common to the south in Sinaloa.

Cercidium floridum Benth. Palo verde, ca'aro (M). The blue palo verde grows along arroyos to more than 10 m tall. It gives comforting shade, even in the depths of the spring drought. Ca'aro is excellent for livestock, who relish the beans and, in times of drought, the green bark. The wood is fashioned into saddle horns.

Cercidium praecox (S. Watson) Sarg. Brea, choy (M). A green-barked, bushy tree with numerous scraggly branches. It has a longer trunk and longer, more irregular branches than *C. floridum*. It flowers in early spring. The persistent pods of the ripe fruits give the tree an orange hue.

The sap *(chu'uca)* is used as a remedy for diabetes, bronchitis, and asthma and is eaten as a sweet. Vicente Tajia Yocupicio produced several ounces of the sap and offered it to one of us who suffers from asthma, assuring him that the chu'uca would provide relief. Doña Buenaventura Mendoza of Teachive reported that when she was a child the ashes of brea roots were boiled with a pig carcass to produce soap, which, she assured us, was as effective as any soap now available. Animals eat the beans and the bark. The branches are used as firewood. The flowers are gathered during Holy Week and sprinkled on house crosses.

Chloroleucon mangense (Jacq.) Britton & Rose. Palo fierro, uose suctu (M; "lion's claws"). A many-trunked tree that quickly spreads to cover a wide area. The convoluted, maculate trunk produces spectacular mosaic patterns of white, black, and gray splotches. The boles are typically hollow inside, providing shelter for myriad invertebrates, small mammals, and *cachoras* (spiny lizards).

The wood is valued for firewood and for making durable posts. The beans and leafage are eaten by livestock. The tree's spreading habit also provides shade for livestock, though not much for humans because its descending thorny branches deter human access.

Conzattia multiflora (B.L. Rob.) Standl. Palo joso de la sierra, joso (M). A stately tree of the higher foothills, often reaching 20 m or more in height, with a smooth, silvery bole that typically rises with great dignity above the canopy. It is leafless much of the year, its elegant form starkly accentuating the surrounding forest. It flowers a glorious yellow in late June, but leafs out only after the summer rains begin and drops its leaves shortly after the rains end, the first large tree to do so. In September its crown bursts into fruits, whose golden presence is seen from afar. In winter the persistent pods turn orange and add an unusual tint to the winter landscape.

The Mayos and Guarijíos indicate that the tree, although handsome, has no use, perhaps because it is uncommon in the lower elevations of the TDF. Its wood is considered to be inferior, subject to rapid rotting and general deterioration. Robert Bye (Bullock et. al. 1995) states that it is one of two trees yielding commercially valuable lumber from the TDF in Mexico. We include it because it is known for its beauty, and hence has aesthetic value.

Coursetia glandulosa A. Gray. Samo, causamo (M; possibly of Guarijío origin), sa'mo (M). A resilient, slender tree (more commonly a shrub) seldom more than 6 m tall, typically found with several trunks springing from a common root. It flowers when leafless in the spring, the decurved branches supporting a rather sparse harvest of white to lavender blossoms.

The straight, springy wood is excellent for arrows. Baskets are made from the branches, chairs from the limbs and trunk. A lac scale insect (*Tachardiella coursetiae*, Kerridae) that frequents the branches produces a hardened sap called *goma de Sonora*. The gum is cooked with cinnamon and mistletoe gathered from mesquite. It is administered as a cure for diarrhea in children who reportedly contract it from their mother's milk. It is said to be a good remedy for stomach ailments. The tree's sap is also taken for stomach distress.

Diphysa occidentalis Rose. Güiloche, güicobo (M). A many-boled small tree growing on more arid or cutover sites in the forest. It flowers a delicate yellow in early winter.

The splendidly straight trunks, seldom more than 5 cm thick but springy, durable, and extremely dense, are cut to make canes and clubs of unyielding texture that resist aging. The trees may grow thicker but are so sought after that one seldom finds a plant without two or three stumps at its base. Gentry (1942) noted that the term *güiloche* referred to a beating with a club of the same material. Due to their strength and durability, the trunks are widely used as latas to support dirt roofs. They are also cut in considerable numbers to make cross-pieces for the attractive Mayo *talabetes* (stools) whose seats are fashioned from animal hides. These will endure decades of heavy use.

Diphysa suberosa S. Watson. Corcho. A small tree with compact, spindly leafage and dark, deeply fissured bark reminiscent of pine species. It grows only sporadically throughout the Mayo region but is abundant in the localities where it does grow. A specimen was not initially recognized by don Pancho Valenzuela, who mistook it for algorrobo *(Acacia pennatula)*. Upon examining the tree more closely, however, he affirmed that it was corcho because it had no clawlike spines and the bark was soft. The bark is used to make *corchos* (corks) for *bulis*, the gourds used by natives to transport water.

Erythrina flabelliformis Kearney. Coral bean, chilicote, flor de Mayo, pionía, jévero (M). A thick-boled tree with silvery gray (sometimes reddish) bark similar to *Cochlospermum vitifolium* but with a noticeable vertical grain in the bark. The trees seldom exceed 10 m and typically rise in several trunks from a common root. Large, broad, downturned, harmless thorns protrude from the bark. The trunks taper quickly, thus resembling a candle. In May and June the brilliant red spears of flowers can be seen from great distances in the dreary brown and gray of the parched forest, the color referred to locally as *mojino*. In fall and spring the persistent pods may linger, often dehiscing lazily to reveal a flash of bright red beans.

The young seeds are toasted, ground, and brewed into a tea for diarrhea and for the stings of scorpions. The bark is also directly applied to the stings. The thorns are boiled into a tea that is taken for urinary problems. The balsalike wood is made into benches and corks. A bench built of the wood at El Rincón was light and quite comfortable. At Nahuibampo log rafts were once made for crossing the Río Mayo during flood stage. Several consultants asserted that the green beans are toasted and the seeds eaten like peanuts, suggesting that the reputed toxic properties of the red beans amount to North American superstition. In the Sierra de Alamos and near Güirocoba, poles are planted and grow to become a living fence. The root is boiled to produce a yellow dye.

Eysenhardtia orthocarpa (A. Gray) S. Watson. Kidneywood, palo dulce, baijguo (M). Typically a shrub in thornscrub or Sonoran desertscrub, it becomes a tree of considerable size in protected canyons in the Mayo region.

The durable wood, nowhere common, is said to be the hardest of local woods and is used in walls, vigas, and cross-veining, the straight sticks laced on vigas to support the earthen roofs of Mayo houses. The wood is also said to be ideal for tool handles. It is boiled to make a blue dye, which is quite permanent when wool is steeped in it overnight. Palo dulce was said to be capable of curing vomiting in hens when a small piece of the wood was placed in their drinking water. (How often hens vomit was not stated.) Cows are said to get fat on it, a possible explanation for the Spanish *dulce*, meaning "sweet."

Haematoxylum brasiletto Karst. Brasil, júchajco (M). A common small tree seldom growing more than 6 m tall, with a strongly fluted trunk of mottled gray and charcoal color that is often an imposing work of natural artistry. Its uses are many and varied. It is so valuable that life in the TDF would be quite altered without it. Unfortunately, it is heavily harvested and grows slowly.

The red part of the cambium is boiled and made into a tea for high blood pressure and for strengthening the heart but is sometimes drunk purely for its refreshing qualities. The liquid is bottled, and a few drops are used as a cure for *tirisia* (sadness or perhaps depression). The brownish liquid, which becomes reddish with greater concentration and is sometimes mixed with tajimsi *(Krameria erecta)* root, is used as a dye for rugs, masks, and basket materials. Sharpened points are used to shuck corn and are attached to lances wielded during Lenten festivities. The wood is regarded as excellent firewood, which makes the slow-growing trees excessively exploited. The wood also makes corrals of a most agreeable appearance. A *tinajera* (tripod for water jug) at San Antonio was constructed from the wood.

Havardia mexicana (Rose) Britton & Rose. Palo chino, chino (M). A mesquitelike tree that is easily confused with it but readily differentiated by its tinier leaflets and more delicate aspect. In low-lying areas between Alamos and San Bernardo, especially those with silt soils, it is the dominant tree, often exceeding 12 m in height. Its thick trunk with fissured, dark bark often arises irregularly from the base.

The reddish wood is widely used for furniture, masks, wooden bowls, and planks. It may be used for firewood, although it is not the preferred variety. The bark can be used for tanning hides but is inferior to mauto.

Havardia sonorae (S. Watson) Britton & Rose. Jócono, jócona (M). A small leguminous tree with intense spines that grow in clear, definite rings around the trunk and branches. It grows in drier parts of TDF and moister parts of thornscrub and reaches 8 m in height. It produces white globular flowers in June and July. The blossoms produce an intensely sweet perfume, attracting swarms of bees and hummingbirds. The legumes are favored by livestock. Vicente Tajia Yocupicio said that the bark of the jócona is boiled with salt to produce a tea, which is put in a nonplastic cup and drunk to cure a sore throat. The wood is also used as firewood. The trees are viewed with disfavor by ranchers, who, after clearing them, find they quickly grow back in dense thickets with vicious thorns.

Leucaena lanceolata S. Watson. Palo bofo, guaji (M), güique (M). A slender tree with a branchless trunk and a broad crown of lacy, leguminous foliage. It is rather rare, and few individuals have a name for it. Although no specific uses for the tree were described because of its scarcity, don Pancho mentioned that it would probably be useful for fence posts. Vicente Tajia Yocupicio claims it can be used for firewood.

Lonchocarpus hermannii Marco. Nesco (M). A common tree of the forest characterized by a somewhat twisted bole with gray bark that typically invaginates in unusual patterns. Usually a slender tree, it grows to over 10 m tall in mesic locations in TDF or along moist washes in coastal thornscrub. In late May and June it bursts into bloom with a vast spray of lavender to pink flowers that are visible from a distance.

The bark is brewed into a tea and used as a bath for mange on animals and lice on people. It was suggested as a remedy for dandruff. It is also used to stun fish.

Lysiloma divaricatum (Jacq.) J.F. Macbr. Mauto, mayo (M). A stalwart of the TDF, untold numbers of these trees rise to a height of 8 to 10 m. Their anvil-shaped crowns meet and form a nearly uniform canopy over a large area. As the trees mature, small sections of stiff bark peel away, presenting a characteristic rugged appearance. Natives distinguish between mauto blanco, colorado, and prieto, which, they assert, have entirely different woods; these varieties are not readily discernible to botanists.

The wood is greatly valued for fence posts and for firewood. Only amapa and palo colorado are ranked superior for house construction. A tea made from boiling the bark is said to be good for diarrhea. It is purported to cleanse the stomach and help expel phlegm. The bark was formerly the basis of the southern Sonoran tanning industry and is still used in Masiaca, an important leather center in the heart of Mayo country, where large backyard areas of leather shops are devoted to drying strips of mauto bark. The bark is harvested without obvious detriment to the tree, which recovers fully within a year. The tree grows rapidly, not as quickly as güinolo, but much faster than mezquite. The Río Mayo takes its name from the tree; thus the translation for Río Mayo is "Mauto River."

Lysiloma watsonii Rose. Tepeguaje, mach'aguo (M). A large prominent tree with dark fissured bark and an angular, irregular trunk that typically reaches 15 m in height and width. The tree develops an individual appearance and apparently reaches a great age. It superficially resembles mezquite with its dark trunk, but its lacy leaf patterns and more open habit distinguish it from that species. It leafs out in April and May with light green (almost yellow) foliage, standing alone against the silver-brown of the sere landscape. The leaves gradually turn darker green through the summer.

The bark is brewed into a tea for gas, gastritis, and liver problems, a symptom of which is the inability to eat and digest fats. It is also a cure for infertility in women who cannot have children because they are "cold" as a result of having eaten cold things (e.g., ice, lemon, or orange) while undergoing menarche. The hot tea from the bark is believed to restore them to the proper temperature. The tea is also generally prescribed for women's problems. Only the inner bark is used.

The wood is used in building, in making posts (notably gate posts), and in making molds for *piloncillos* (brown sugar cones). It is a heavy and resilient but not particularly hard wood.

Mimosa palmeri Rose. Chopo, cho'opo (M). A small tree with wispy leafage that is well armed with potent thorns. The wood is dense and hard. The tree has become scarce because of its renown for charcoal making and fence post durability. It is currently accorded protected status by goverment fiat, which is largely ineffective, if piles of fresh-cut posts in various locations are any indicator. Its thorns are an effective deterrent to close examination but fall readily from the harvester's machete.

Olneya tesota A. Gray. Palo fierro, ejéa (M). An inconspicuous tree in the foothills thornscrub of Mayo country. It reaches the southern limit of its range near Teachive, where it thrives only in a bushy habit in open dry areas and seldom grows more than 7 m in height. To the west, however, the tree becomes a dominant in the coastal thornscrub. Some of the largest specimens, which exceed 10 m in height and width, grow within the confines of the Mayo village of Sinahuisa, a haven from woodcutters and livestock. These individuals are highly valued for the shade they provide. The well-armed trees blossom in late spring, adding a delicate pink lavender to the withering, deciduated landscape. The dead trunk of an enormous ejéa stands protected as a monument in the plaza of Masiaca.

The hard wood is carved into items of personal adornment, such as charms for necklaces and medallions. The wood is used for horcones and firewood. Goats relish the beans. Because of its relative scarcity and the availability of many varieties of wood, it is less significant in the lives of Mayos living in foothills thornscrub than in those of indigenous peoples living in drier habitats to the west and northwest.

Parkinsonia aculeata L. Mexican palo verde, guacaporo (M), bacaporo (M). With age these normally green-barked trees may develop a trunk with black bark, belying the common name. Individuals 10 m in height are not uncommon. The bark or leaves are brewed into a tea for cough. Livestock consume the beans and long, narrow leaves.

Piscidia mollis Rose. Palo blanco, jo'opo (M). A ragged-trunked tree with large oval leaflets resembling those of a grayish oak. Some individuals grow to 10 m in height; trunks in excess of 40 cm in diameter are not uncommon. Its roundish, gray leaves distinguish it from most other trees in the region. The tree often retains its leaves well into the spring drought, all the while appearing craggy and laden with age. Its spreading habit provides welcome shade in the hot sun.

The strong wood is used as a base for porch posts, for the posts themselves, and for beams. The flowers are brewed into a tea and drunk for epilepsy. A tea from the bark is used to stun fish; hence, the generic name *Piscidia*, which means "fish killer."

Pithecellobium dulce (Roxb.) Benth. Guamúchil, maco'tchini (M). A venerable legume highly prized for its fruit and shade. It typically grows to great heights, exceeding 15 m along streams in arroyos. It sometimes succumbs to invasion by strangler figs. Rings of small thorns on the gray bole help identify this excellent tree. It is common in and near villages. Considerable speculation surrounds the geographical origin of the tree because it is found only near human habitation.

The legumes are gathered and relished, often dried as a sweet for later use. Sautéing in a frying pan is said to sweeten them and make them easier to digest. The bark is made into a tea for diarrhea. It is also used in curing hides but is inferior to the bark of mauto because of its strong odor.

Platymiscium trifoliolatum Benth. Tampicerán (M). A straight-boled tree that grows to 15 m in height and produces bright yellow flowers in March. It is found in moist riparian areas above 400 m. When its large leaves fall, usually in spring, the ground below becomes covered with a yellow leafy carpet.

Although uncommon in Mayo country, plants at Rancho El Guayabo on the Río Cuchujaqui were known to don Pancho. The wood is valued for making furniture and buildings. At Jurinabo, it was the preferred wood for making molds for brown sugar loaves. Near Tepopa, it was used in making gate posts. Don Pancho was familiar with these uses, although the producers were non-Mayos.

Prosopis glandulosa Torr. Honey mesquite, mezquite, juupa (M). A versatile tree that grows to great proportions in the TDF along arroyos and in areas where human clearings have permitted seedlings enough sunlight to survive. Individuals may reach 12 m or more in height with an equal spread, and trunks may measure 1 m in diameter.

The wood is used in all phases of building and in the construction of corrals, fence posts, and utensils. It is esteemed as a firewood and is consequently much persecuted. Although its virtues as a shade tree are widely acknowledged, the desire to exploit the wood often outweighs the need for shade, and the tree is sacrificed. The beans (*péchitas*) are sought after by livestock and were formerly eaten by humans as well. The beans are said to be less tasty than those of *P. velutina*, which is uncommon in the Mayo region. The gum (called *chúcata* in Mayo) is kneaded into a ball, mixed with sugar, roasted, and eaten as candy. The sap of the root is used as a dye.

Senna atomaria (L.) Irwin & Barneby. Palo zorrillo, jupachumi (M). A common TDF plant that rises to 10 m. This spreading tree with large, dark green, roundish leaves has lengthy, persistent beans that typically dangle for many months from the point of attachment. The dark gray to black trunk is often blotched with lighter gray or white and in older individuals is dented with slots. Jupachumi flowers a dense yellow on leafless plants in April. The Mayo name means "skunk's ass."

The wood is used for vigas and horcones in house building, for fence posts, and for firewood. The beans are eaten by livestock as they fall to the ground. The beans are boiled and drunk as a remedy for diabetes.

MALPIGHIACEAE

Malpighia emarginata DC. Granadilla, sire (M). A large shrub or small tree seldom more than 4 m tall, with gnarled trunk and maculate, flaking bark. It often branches densely near the base and forms multiple arches and irregular trunks and branches. The tree's peculiar growth pattern bestows an appearance of great age. It grows in the drier areas of the Mayo region. The fruits are eaten and become available during *las aguas*, the summer rains, or earlier in some locations.

MELIACEAE

Trichilia hirta L. Bola colorada, síquiri tájcara (M). A tropical chinaberry, this handsome, spreading, small tree seldom reaches more than 5 m in height. The parallel leaflets on the long rachis resemble the configuration of walnut leaves, and the green fruit superficially resembles developing walnuts. Luís Valenzuela Moroyoqui of Las Rastras identified the tree and believed it had uses, but he could not recall them.

MORACEAE

Ficus cotinifolia H.B.K. Strangler fig, nacapuli (M). The great fig typically grows to immense size—more than 20 m in height and width—and creates massive entanglings of roots. The fruits are relished by humans, beasts, birds, and insects. The tree is greatly valued for shade, a blessing in the stifling heat of late spring and summer. The name is derived from the Mayo *naca* (ear).

Ficus insipida Willd. Chalate, tchuna (M). This fig produces the best fruits, according to Mayos. Some natives will travel considerable distances to gather the succulent figs. Its leaves are large and pointed.

Ficus petiolaris H.B.K. Rock fig, tescalama, báisaguo (M). Found on more arid slopes and watercourses, its yellow-white bark and large, dark green leaves are distinctive. Its roots assume grotesque shapes and may cover great areas, cascading down rocks like a frozen yellow waterfall. Though it is a smaller fig, individuals may attain heights in excess of 15 m.

The sap is rubbed on a rag or the bark is wrapped in a rag; both methods are used to bind hernias. It is used the same way to correct dislocations and sprains. Vicente Tajia Yocupicio collected a bagful in Arroyo Las Rastras, some 20 km from his home, and took it home to replenish his pharmacopoeia.

Ficus trigonata L. Higuera, nacapuli (M). A huge fig tree that produces abundant shade as well as immense numbers of small fruits, to the delight of numerous birds, insects, and mammals.

Nyctaginaceae

Pisonia capitata (S. Watson) Standl. Garambullo, baijuo (M). An unpredictable plant braced heavily with enormous stern thorns and hooked spines. It often grows into vast forbidding thickets that are impenetrable even to cattle; less often, it grows as a straight-trunked tree. The intimidating spines, resembling ice picks with heavy hooked thorns, are said by natives to be poisonous. Baijuo is a most unfriendly plant.

It is planted to make a living fence that excludes all but the smallest creatures and such *culebras* as the brown vinesnake *(Oxybelis aeneus)*, indigo snake or bacomuteca ([M]; *Drymarchon corais)*, parrot snake *(Leptophis diplotropis)*, and Sonoran red racer *(Masticophis flagellum* subsp. *cingulum)*. The wood is carved into masks and musical instruments.

Olacaeae

Schoepfia schreberi J.F. Gmel. Palo cachora, cuta béjori (M), júchica (M). A symmetrical tree that is typically small but grows to nearly 10 m in the Sierra de Alamos. Its leaves are small, crinkly, and superficially similar to oak leaves. *Cachora* refers to lizards such as *Ctenosaura hemilopha* (spiny-tailed iguana) and *Sceloporus clarki* (Clark's spiny lizard), which often live inside the hollow trunks. Mayos appear to hesitate to ascribe uses to this tree, although they acknowledge its possible value for fence posts.

Opiliaceae

Agonandra racemosa (DC.) Standl. Palo verde, matachamaco, úsim yuera (M). A tall, usually slender tree commonly found along watercourses in the Mayo region. Some individuals grow into large trees that are taller than 10 m. This species bears shiny, dark to bright green, oval leaves that tend to droop. The wood is used in all phases of house construction. The leaf is cooked as a cure for snake bite.

Palmae (Arecaceae)

Brahea aculeata (Brandegee) Moore. Palma, ta'aco (M). A small palm whose leaves, which are harvested when quite green, are valued for making baskets. It is also a standard roofing material for homes, providing unmatched protection from rain and heat. Few building materials present a more agreeable appearance than the palm thatching used in many older Mayo homes in the TDF. Unfortunately, it is not found in the lower parts of the region, and hence, is not available for homes there.

Rhamnaceae

Karwinskia humboldtiana (Roem. & Sch.) Zucc. Cacachila, aroyoguo (M). Normally a leafy shrub, on occasion it grows into a tree as tall as 10 m. The leaves are a most agreeable, smooth, gray-green color, typically with black dashes along the veins on the underside.

A tea is made from the leaves for liver problems. According to Vicente Tajia Yocupicio, when your eyes become yellow and your urine very yellow, you have hepatitis and should use the tea. For colds, the leaves are steeped in alcohol during the rainy season. Thereafter, the liquid is stored in a vial and inhaled. The wood is used for posts in houses.

Ziziphus amole (Sessé & Moc.) M.C. Johnst. Saituna, baís cápora (M). A spiny tree with large, dark green leaves. In coastal thornscrub, where it stands out among the columnar cacti, reaching a height of more than 10 m, its branches often become laden with ball moss (*Tillandsia recurvata*), which is valued as livestock feed. The fruits are eaten. The bark is made into a tea taken to expel amoebas. The leaves are a principal source of fodder for goats. Vicente Tajia Yocupicio reported that the Mayo name means "three old corporals."

RUBIACEAE

Hintonia latiflora (Sessé & Moc.) Bullock. Copalquín, tapichogua (M). A highly esteemed tree up to 10 m tall, with rough brownish bark (often one side of the tree has smooth bark as well) and a dense crown of leaves growing in whorls toward the tips of the branches. In September the pale green, club-shaped buds with six sharp ribs and the ivory-colored, trumpet-shaped flowers that hang pendulous from the tips of branches present a most odd appearance. When the tree is leafless in the early spring, the walnut-sized fruits persist, assisting in identification.

The bitter-tasting bark is made into a tea for numerous ailments, including malaria. To thicken the blood or alleviate anemia, the tea is drunk or the bark is chewed directly from the tree. The bark is widely believed to have the power to kill microbes in the blood. For cough, it is viewed as an excellent remedy. It is thought to help control diabetes. An individual in Teachive keeps a bundle of it in his house and drinks the bitter tea daily to help combat his diabetes. The bark is also ground into a powder and applied to chigger bites. The wood is used for vigas in houses. Some trees are thought to be more potent than others and show scarring from repeated harvesting of bark. Ground copalquín is sold as a remedy in some local markets.

Randia echinocarpa Sessé & Moc. ex DC. Papache, jósoina (M). A scraggly, typically arching shrub or small tree with prominent thorns often more than 5 cm in length. It bears a strange, tough-skinned fruit the size of tennis balls with large harmless thorns protruding from the surface. When the fruits mature, the pulp becomes a black mass quite popular with natives as a food. When eaten on an empty stomach, the pulp is said to help expel amoebas.

RUTACEAE

Casimiroa edulis La Llave & Lex. Chapote, zapote. A majestic spreading tree that typically grows to 15 m. It is valued for shade and is often seen near human habitation, suggesting that its origin in Sonora may be connected with human migrations. The fruits are said to be edible. Judging from the hordes of honeybees visiting the female flowers, we surmise that it is an important source of honey.

Esenbeckia hartmani B.L. Rob. & Fernald. Palo amarillo, jójona (M), momoguo (M). A small tree, often many trunked, that grows inconspicuously in the forest. When the branches are broken, the wood appears yellowish; hence, the Spanish name. The wood is sometimes used for posts and buildings.

SALICACEAE

Populus mexicana Sarg. subsp. *dimorpha* (Brandegee) Eckenwalder. Alamo, aba'aso (M). A tall (to 20 m), spreading tree that grows where groundwater is available. For blows and bruises, the bark is cooked with salt and the wound is washed with the liquid. The wood of the extensive root is used to carve figurines, masks, and household implements. It is also sawed into planks for tables. The freshly leafed-out branches are used as adornment in Holy Week festivities.

Salix bonplandiana H.B.K. Sauce, huata (M). (Huatabampo = "willow in the water.") A medium-sized willow growing in the Río Cuchujaqui. Don Pancho Valenzuela believed it had uses but was unfamiliar with them. At El Mesquital near Navojoa, the springy branches are woven into baskets.

Sapindaceae

Sapindus saponaria L. Abolillo, amolillo, tubchi (M). A common tree that becomes quite large when it grows near water, reaching 12 m in height and nearly as much in width. The wood is made into crucifixes and beads for rosaries and necklaces. The black seeds of the fruit are eaten or strung into attractive necklaces worn with pride by Mayo *fiesteros* (fiesta leaders). The fruits are made into soap.

Sapotaceae

Sideroxylon occidentale (Hemsl.) Pennington. Júchica (M). A rare tree at Huasaguari that rises to 12 m, with a checkered bark and *Lycium*-like leaves. The tree is common at Teachive but has not been found in other locations. The wood is carved into dishes and utensils.

Sideroxylon persimile (Hemsl.) subsp. *sessiflora* (Hemsl.) Pennington. Bebelama (M). A large spreading tree found only in moist, deep canyons. It provides cool shade in Arroyo las Bebelamas near El Rincón. Don Pancho noted that the fruits are edible once they are cooked.

Sideroxylon tepicense (Standl.) Pennington. Tempisque, ca'ja (M). A tall spreading tree up to 20 m tall, with a star-shaped, palmate leaf pattern. Knowledge of the locations of trees is maintained as family tradition.

The fruits are relished. Some say they should be cooked because they tend to burn the lips and mouth when eaten raw. The bark is boiled, and the resultant tea is drunk for fever. The bark can be used in place of rennet for coagulating cheese.

Simaroubaceae

Alvaradoa amorphoides Liebm. Palo torsal, guaji (M), paenepa (M). A small, uncommon tree up to 10 m high, with spreading branches and a lengthy (10-cm) inflorescence of many reddish flowers, which bloom in early winter. In flower it is a most handsome tree. It appears to be limited in range to moister regions of the TDF, mainly in canyons. Don Pancho concurred with its common name and said it has uses, but he could not recall what they were.

Sterculiaceae

Guazuma ulmifolia Lam. Guásima, agia (M). A common, spreading, elmlike tree in riparian habitats in TDF that grows as high as 12 m. It often retains its leaves through the heat of spring, even though they may curl up and be practically dead.

Guásima is the wood of choice for constructing movable furniture: chairs, tables, cribs, and cradles. The color of the wood varies from a handsome white to a delicate pink. The time of appropriate harvest is critical, coinciding with the full moon; September is the best month. Furniture of guásima wood harvested in a timely fashion will last more than 40 years. If harvested at the wrong time, the wood will crack, be less strong, be subject to boring insects, and fail within a single generation, or so natives claim. Natives also claim that branches or boles from which lumber has been harvested will regenerate in 2 years. The inner bark is boiled and made into a tea for liver sickness. Strips of the bark are chewed for the sweet flavor and as a tonic. The dried fruits are ground and brewed for a coffee substitute. Agia is a most important tree in the lives of the Mayos.

TAXODIACEAE

Taxodium distichum (L.) J.M.C. Rich. var. *mexicanum* Gordon. Mexican bald cypress, sabino. The tree is well known in the region, but we have not determined a Cáhita name for it. Sabinos grow at the edge of permanent water in watercourses, primarily the Río Cuchujaqui in the Alamos area but also along Arroyo El Tábelo and Arroyo Techobampo to the north. They become massive, often exceeding 25 m in height, and their roots form an intricate, braided mass. In spring they replace their old needles with bright green herbage, adding a touch of freshness to a dying landscape. The tree's great roots mesh powerfully into bedrock, grounding them against flash flooding. During the massive floods in the Río Cuchujaqui in the wake of Hurricane Ismael in September 1995, many huge sabinos were downed and swept away by the powerful currents that reached a depth of 8 m.

The aromatic lumber of the sabino is indisputably the best wood for making permanent tables and other furniture. The trees are protected by the Mexican government, a measure that seems to be effective. Natives living along the lower Río Cuchujaqui occasionally encounter sabino driftwood in the meandering sands of the watercourse. With the help of beasts of burden, they haul these logs to their homes and guard them as treasures. One consultant had stored a ponderous log for several years and was still contemplating using it to make some article of furniture.

THEOPHRASTACEAE

Jacquinia macrocarpa Cav. subsp. *pungens* (A. Gray) B. Ståhl. Sanjuanico, tásiro (M). A small tree, often symmetrical in outline, with a typically thick trunk bearing smooth gray bark. The small (less than 3 cm long), green, persistent, stiff leaves are ovate with a needle-sharp tip capable of puncturing the skin without pain when gently probed. If one approaches the herbage too quickly, however, numerous skin punctures can ensue with disagreeable results.

The mature fruits, which are the size of large grapes, are mashed in water, allowed to stand, and then used as a shampoo. The dried orange flowers are strung into necklaces and open to their original shape when soaked. The dried flowers are boiled with wool to produce a yellow dye of considerable beauty. They must be harvested before las aguas or the dye will run. One consultant stated with a straight face (and others corroborated) that the leaves are mashed into a ball and inserted into the anus of a constipated cow for immediate relief to the cow. Doña Buenaventura Mendoza (born 1905) of Teachive regularly collects the flowers and brews them into a tea, which she drinks to strengthen her heart.

TILIACEAE

Heliocarpus attenuatus S. Watson. Samo baboso, sa'amo (M). A slender springy tree up to 7 m tall with small, white terminal flowers and large, heart-shaped leaves. The dried flowers persist, turning brown and resembling tiny images of the sun; hence, the generic name means "sun fruit." Pieces of the bark are soaked in water and mixed with lime to make whitewash adhere to plaster. If it is not added, the whitewash will slowly chip away, producing a blotched color. Long strips of the bark are used for lashing poles together. They are reported to shrink upon drying, making a secure binding. The bark is also given to constipated animals. *Baboso* means "drooling" in Spanish, referring to the foam the bark makes when soaked in water.

ULMACEAE

Celtis iguanea (Jacq.) Sarg. Tropical hackberry, vainoro, cumbro (M). Vainoro grows in great abundance, usually producing huge, arching branches armed with sharp spikelike thorns up to 12 cm in length. These branches often grow horizontally, finally curving downward to reach the earth and form an impenetrable barrier. The tree often combines with *Pisonia capitata* to form thickets along watercourses that inspire impolite oaths from passersby. Trees will assume a tangled appearance in which other branches and vines are often intermingled, greatly confusing the onlooker. The cranberry-sized orange fruits are eaten with great relish by birds and humans, even though the large seed occupies most of the fruit. The wood is used for posts, handles, and roof cross-hatching.

Celtis pallida Torr. Desert hackberry, cumbro (M). Mayos declared the name *cumbro* to be Mayo, but the origin may be *cúmero*, the Spanish name for hackberry. Or perhaps *cúmero* is derived from *cumbro*. The fruits of cumbro are relished. The roots are made into a tea used as a purgative.

Celtis reticulata Torr. Netleaf hackberry, aceituna, cúmero, cumbro (M). In the Mayo country, this normally rough-barked tree grows a straight trunk with smooth bark up to 10 m tall, assuming a symmetry unusual in more temperate areas to the north. The bark is boiled into a tea and taken for tonsillitis. The orange fruits are eaten, yielding a sweet taste seldom experienced because of the sparse and diminutive nature of the fruits.

VERBENACEAE

Vitex mollis H.B.K. Igualama, uvalama, júvare (M). A spreading tree that sometimes grows to 15 m. In April it flowers delicately, its lavender blooms exuding a perfume quite irresistible to pollinators. The black fleshy fruits are widely eaten, and the leaves are relished by livestock.

Vitex pyramidata Rob. Negrito, ju'upa'ari (M). A small tree with clasping, folding leaves, found in more arid habitats than those of the igualama, especially on acidic or ash soils. The fruits, though smaller than those of the igualama, are equally relished.

ZYGOPHYLLACEAE

Guaiacum coulteri A. Gray. Guayacán, júyaguo (M). A large shrub in the Sonoran Desert and thornscrub, which becomes a tree of greatly varied habit, often with a thick, powerful trunk that grows to 6 m in the TDF. The tiny, brilliant, emerald green leaves, which are adnate to the branches, shine through the forest in the dry months. In April the dark purple-blue, fragrant blossoms attract pollinators and humans. Some of the trees are valued for the shade they bestow during the hot months.

The flower is cooked, and the tea is drunk for asthma. The heartwood is brewed into a tea to cure *pujos* (bloody stools). The hard, strong wood is used for fence posts. The smoke is reported to be toxic, causing the hair of those who inhale it to fall out.

Appendix 4.2. Trees of the Mayo Region by Scientific, Mayo, and Spanish Names

Scientific Name	Mayo Name	Spanish Name
Acacia cochliacantha	Chírajo	Chírahui, güinolo
Acacia coulteri	Baigüío	Guayavillo
Acacia farnesiana	Cu'uca	Vinorama
Acacia occidentalis	Teso	Teso
Agonandra racemosa	Úsim yuera	Matachamaco, palo verde
Albizia sinaloensis	Joso	Palo joso
Alvaradoa amorphoides	Guaji, paenepa	Palo torsal
Brahea aculeata	Ta'aco	Palma
Brongniartia alamosana	Cuta nagüía	Palo piojo, vara prieta
Bursera fagaroides	To'oro, to'oro saguali	Torote, torote de vaca, torote de venado
Bursera grandifolia	To'oro mulato	Palo mulato
Bursera lancifolia	To'oro chutama	Torote copal
Bursera laxiflora	To'oro chucuri	Torote prieto
Bursera penicillata	To'oro	Torote acensio, torote de incienso
Bursera stenophylla	To'oro, to'oro sajo	Torote copal
Caesalpinia caladenia	Jícamuchi	Palo piojo blanco
Caesalpinia platyloba	Cuta síquiri, mapáo	Palo colorado
Caesalpinia sclerocarpa	Ebano	Ebano
Carnegiea gigantea	Saguo	Saguaro
Casimiroa edulis		Chapote, zapote
Ceiba acuminata	Baogua	Pochote
Celtis iguanea	Cumbro	Vainoro
Celtis pallida	Cumbro	
Celtis reticulata	Cumbro	Aceituna, cúmero
Cercidium floridum	Ca'aro	Palo verde
Cercidium praecox	Choy	Brea
Chloroleucon mangense	Uose suctu	Palo fierro
Cochlospermum vitifolium	Ciánori	Palo barriga, palo barril
Conzattia multiflora	Joso	Palo joso de la sierra
Cordia sonorae	Pómajo	Palo de asta
Coursetia glandulosa	Causamo, sa'mo	Samo
Croton fantzianus	Cuta tósari	Vara blanca
Croton flavescens	Júsairo	Vara prieta
Diospyros sonorae	Caguorara	Guayparín
Diphysa occidentalis	Güicobo	Güiloche

Scientific Name	Mayo Name	Spanish Name
Diphysa suberosa		Corcho
Erythrina flabelliformis	Jévero	Chilicote, flor de Mayo, pionía
Erythroxylon mexicanum	Mamoa	Mamoa
Esenbeckia hartmani	Jójona, momoguo	Palo amarillo
Eysenhardtia orthocarpa	Baijguo	Palo dulce
Ficus cotinifolia	Nacapuli	Nacapuli
Ficus insipida	Tchuna	Chalate
Ficus petiolaris	Báisaguo	Tescalama
Ficus trigonata	Nacapuli	Higuera
Forchhammeria watsonii	Jito	Palo jito
Fouquieria macdougalii	Murue	Jaboncillo, ocotillo, palo pitillo
Guaiacum coulteri	Júyaguo	Guayacán
Guazuma ulmifolia	Agia	Guásima
Haematoxylum brasiletto	Júchajco	Brasil
Havardia mexicana	Chino	Palo chino
Havardia sonorae	Jócona	Jócono
Heliocarpus attenuatus	Sa'amo	Samo baboso
Hintonia latiflora	Tapichogua	Copalquín
Ipomoea arborescens	Jútuguo	Palo santo
Jacquinia macrocarpa	Tásiro	Sanjuanico
Jatropha cordata	To'oro	Panalero, torote papelío
Jatropha malacophylla	Sa'apo	Sangrengado
Karwinskia humboldtiana	Aroyoguo	Cacachila
Leucaena lanceolata	Guaji, güique	Palo bofo
Lonchocarpus hermannii	Nesco	Nesco
Lysiloma divaricatum	Mayo	Mauto
Lysiloma watsonii	Mach'aguo	Tepeguaje
Malpighia emarginata	Sire	Granadilla
Mimosa palmeri	Cho'opo	Chopo
Montanoa rosei	Batayaqui	Batayaqui
Olneya tesota	Ejéa	Palo fierro
Pachycereus pecten-aboriginum	Etcho	Etcho
Parkinsonia aculeata	Bacaporo, guacaporo	Bacaporo, guacaporo
Piscidia mollis	Jo'opo	Palo blanco
Pisonia capitata	Baijuo	Garambullo
Pithecellobium dulce	Maco'tchini	Guamúchil
Platymiscium trifoliolatum	Tampicerán	
Plumeria rubra		Cascalosúchil
Populus mexicana	Aba'aso	Alamo
Prosopis glandulosa	Juupa	Mezquite
Pseudobombax palmeri	Cuajilote	
Quercus chihuahuensis	Encino	

Appendix 4.2. (Continued)

Scientific Name	Mayo Name	Spanish Name
Randia echinocarpa	Jósoina	Papache
Salix bonplandiana	Huata	Sauce
Sapindus saponaria	Tubchi	Abolillo, amolillo
Schoepfia parviflora	Cuta béjori	Palo cachora
Schoepfia schreberi	Cuta béjori, júchica	Palo cachora
Sebastiania pavoniana	Túbucti	Brincador
Senna atomaria	Jupachumi	Palo zorrillo
Sideroxylon occidentale	Júchica	
Sideroxylon persimile	Bebelama	Bebelama
Sideroxylon tepicense	Ca'ja	Tempisque
Stenocereus montanus	Sahuira	Sahuira
Stenocereus thurberi	Aaqui	Pitahaya
Tabebuia chrysantha	To'obo saguali	Amapa amarilla
Tabebuia impetiginosa	To'obo	Amapa
Taxodium distichum	Sabino	Sabino
Tecoma stans		Gloria
Trichilia hirta	Síquiri tájcara	Bola colorada
Vallesia glabra	Sitavaro	Sitavaro
Vitex mollis	Júvare	Igualama, uvalama
Vitex pyramidata	Ju'upa'ari	Negrito
Wimmeria mexicana	Chi'ini	Algodoncillo
Ziziphus amole	Baís cápora	Saituna

REFERENCES

Acosta, R. 1949. *Apuntes históricos Sonorenses*. Hermosillo: Gobierno del Estado de Sonora.

Bullock, S. H., H. A. Mooney, and E. Medina, eds. 1995. *Seasonally dry tropical forests*. Cambridge: Cambridge University Press.

Camou H., E. 1985. Yaquis y Mayos: Cultivadores de los valles. In *Historia general de Sonora*, vol. 5, ed. S. Calderón Valdés, 299–301. Hermosillo: Gobierno del Estado de Sonora.

Crumrine, R. 1983. Mayo. In *Handbook of North American Indians*, vol. 10, ed. A. Ortiz, 264–275. Washington, D.C.: Smithsonian Institution.

Felger, R. S., and M. B. Moser. 1985. *People of the desert and sea: Ethnobotany of the Seri Indians*. Tucson: University of Arizona Press.

Gentry, H. S. 1942. *Río Mayo plants: A study of the flora and vegetation of the valley of the Río Mayo, Sonora*. Publication No. 527. Washington, D.C.: Carnegie Institution of Washington.

———. 1963. The Warihío Indians of Sonora-Chihuahua: An ethnographic survey. Anthropological Papers No. 65. *Bureau of American Ethnology Bulletin* 186:61–144.

Hastings, J. R., and R. H. Humphrey. 1969. *Climatological data and statistics for Sonora and northern Sinaloa*. Technical Reports on the Meteorology and Climatology of Arid Regions No. 19. Tucson: University of Arizona Institute of Atmospheric Physics.

López-Estudillo, R. A. 1993. Contribución a la etnobotánica de Sonora: Las plantas útiles de los Mayos del municipio de Etchojoa, Sonora. Tesis de Licenciatura, Universidad Autónoma de Nuevo Leon, Monterrey.

O'Connor, M. I. 1989. *Descendants of Totoliguoqui*. Berkeley: University of California Press.

Spicer, E. H. 1962. *Cycles of conquest: The impact of Spain, Mexico, and the United States on Indians of the Southwest, 1533–1960*. Tucson: University of Arizona Press.

———. 1980. *The Yaquis: A cultural history*. Tucson: University of Arizona Press.

5

CROP DIVERSITY
AMONG INDIGENOUS FARMING
CULTURES IN THE TROPICAL
DECIDUOUS FOREST

Barney T. Burns, Mahina Drees, Gary P. Nabhan,
and Suzanne C. Nelson

Rising from 400 m above sea level on the coastal plains of the Sea of Cortez to more than 2,500 m in the Sierra Mohinara, the northern Sierra Madre Occidental is a region of tremendous habitat heterogeneity and one of the world's centers of megadiversity for plants (Felger et al. 1997). The region's plant diversity is nested within the eight major physiognomic vegetation types represented in the 180,000 km² found between the northernmost mountains in this cordillera near the Chihuahuan border with New Mexico and the *barranca* (ravine)–riddled highlands on the Sinaloa-Durango border. These vegetation types include mixed-conifer forest, pine-oak forest, oak forest, oak woodland, grassland, tropical deciduous forest, thornscrub, and Sonoran desertscrub (see Van Devender et al., this volume). Within this vegetation mosaic, botanists have already recorded at least 2,500 species of vascular plants (Bye 1995) and project that perhaps as many as 4,000 plant species thrive within the regions defined here (Felger et al. 1997). Although it remains unclear how many of these species are found within the tropical deciduous forest (TDF) alone, it is known that TDF habitats include many of the region's 250 endemic species (Felger et al. 1997).

The habitat heterogeneity of the northern Sierra Madre Occidental has obviously harbored a remarkable level of wild plant diversity, but there is also a great diversity of cultivated plants within the region. Crop land races, known in Mexico as *razas criollas*, are often specifically adapted to particular microclimates and peculiar agronomic conditions instead of being bred for wide adaptability. We hypothesize that the natural habitat heterogeneity of the region is one of several factors that has favored native crop diversification in the region. Wild plant diversity and habitat heterogeneity, in turn, have presumably been shaped by the steep elevation and precipitation gradients in the sierra; by the geological mixture of Cretaceous and Tertiary granites, Tertiary-Quaternary basalts and ignimbrites, and Miocene conglomerates; by the antiquity of the Madro-Tertiary flora; and by the relatively frost-free conditions at lower elevations, particularly in the TDF (Bye 1995; Felger et al. 1997; Nabhan 1990; Wiseman 1980).

However, it remains unclear whether or not these same factors have been similarly influential on the cultivated plant diversity of the region. Other factors to consider are the region's agrohabitat heterogeneity; its cultural-linguistic diversity and occupational history; introgression with wild crop relatives; and the slow, uneven penetration of modern agricultural development, which has more rapidly stripped other regions of their former diversity. Although it is difficult if not impossible to quantitatively measure the relative contributions of these factors in fostering native crop diversity, we wish to qualitatively evaluate how each comes into play. Our focus will be on crop diversity in the TDF zone, but because crop diffusion across zones has affected crop distribution and evolution, we will keep as a necessary backdrop the dynamics of intermixing and isolation in the northern Sierra Madre Occidental as a whole. Our results are based on interpretation of the inventory of native crops that the staff members and collaborators of Native Seeds/SEARCH (NS/S) and the Arizona-Sonora Desert Museum have collected from the sierra since 1976. This inventory has become the largest collection of New World crops from any region north of the Tropics. We will pay particular attention to those species and *land races* (folk varieties) that are unique to the TDF zone.

Aridoamerican Center of Crop Diversity and the TDF Crop Assemblage

The northern Sierra Madre Occidental as a whole has the greatest species richness of pre-Columbian domesticated crops of any region in the Americas north of the tropic of Cancer (table 5.1). There are more native crop species among indigenous farmers in the northern Sierra Madre Occidental than among each of these groups: farmers on the coastal plains of Sonora and Sinaloa; Piman and Yuman farmers on the Gila and Colorado Rivers; upland

Table 5.1. Native Wild and Domesticated Crops of the Northern Sierra Madre Occidental

Crop Species/Race	Presence in TDF Zone [a]	NS/S Collections	Published References & Personal Observations
		Presence among Indigenous Groups [b]	
Agave angustifolia	+		GU, LP, MA, TA
Amaranthus cruentus	+	GU, MA, MP, NT, TA	
Amaranthus hypochondriacus	+	GU, NT	MA
Brassica campestris	o	TA	
Capsicum annuum	+	GU, LP, MA, MP, NT, OP, TA	YA
Chenopodium berlandieri	+	MA	
Cucurbita argyrosperma	+	LP, MA, MP, OP, TA, YA	GU
Cucurbita ficifolia	o	NT, TA	MP
Cucurbita moschata	+	GU, LP, MA, OP, YA	
Cucurbita pepo	o	MP, NT, TA	
Helianthus annuus	o	TA	
Hyptis suaveolens	+	GU, LP, MA, YA	
Jaltomata procumbens	o		TA
Lagenaria siceraria	+	GU, MA, NT, TA, YA	LP, MP, OP
Nicotiana rustica	+	GU, MA, MP, NT, TA	YA
Nicotiana tabacum	+	GU, MP	MA, NT, TA, YA
Panicum sonorum	+	MA	GU
Phaseolus acutifolius	+	GU, MA, OP, TA, YA	
Phaseolus coccineus	o	TA	NT
Phaseolus salicifolius	+		NT
Phaseolus vulgaris	+	GU, MA, MP, NT, OP, TA, YA	LP
Zea mays			
Apachito	o	MP, NT, TA	
Chapalote	+	MA	TA
Dulcillo del noroeste	+	GU, MA, MP, NT, TA, YA	
Gordo	o	TA	
Maiz azul	o	GU, MP, NT, TA	
Onaveño/blando de Sonora	+	GU, MA, MP, TA	LP, OP
Reventador	+	YA	LP, OP
Tabloncillo	+	MA	MP
Tuxpeño norteño	+	GU, MA, TA, YA	

Sources: Felger et al. (1997), Gentry (1963), Merrick and Nabhan (1984), Nabhan (1979, 1984, 1985b, 1990), Ortega Paczka (1985), and Pennington (1963, 1969, 1970, 1982, 1989).

[a] + = present in tropical deciduous forest (TDF) zone; o = absent from TDF zone.

[b] NS/S = Native Seeds/SEARCH. Codes for tribal names: GU = Guarijío; LP = Lowland Pima; MA = Mayo; MP = Mountain Pima; NT = Northern Tepehuán; OP = Opata/Eudeve descendants; TA = Tarahumara; YA = Yaqui.

Yuman farmers in the middle Colorado watershed; Pueblo farmers on the Rio Grande and Hopi farmers in northern Arizona; or farmers in the upper Missouri watershed and the southeastern United States. North of the Valley of Mexico, only the Huastec Mayan crop assemblage is comparably diverse (Alcorn 1984).

Both species and land races found in the northern Sierra Madre Occidental are more characteristic of the Aridoamerican center of crop diversity than of the better-known Mesoamerican center (Nabhan 1985b). For instance, Sonoran panicgrass *(Panicum sonorum)* is shared between the Sierra Madre Occidental and the lower Colorado River tribes (Nabhan and de Wet 1984), but not with Mesoamerican farmers, as is often cited. Similarly, Mesoamerica is often cited as the center of origin and diversity for tepary beans *(Phaseolus acutifolius)*, yet they are much more strongly associated with the Sonoran Desert floodplain farmers and Sierra Madre horticulturalists in eastern Sonora and Chihuahua than they are with Mesoamerica. These beans are occasionally grown in the more humid climes of southern Mexico and Guatemala, but nowhere have they had the prehistoric or historical importance they have had in drought-susceptible Aridoamerica (Nabhan and Felger 1978).

The following crop-by-crop survey highlights the relative importance, land race diversity, and native nomenclature for domesticated species found in the TDF. As described by Wiseman (1980), three related vegetation subzones predominate in southeastern Sonora, adjacent Chihuahua, and Sinaloa: the Tropical Deciduous Scrub Forest, from 28°N to 26°N latitude; the TDF Northern Phase, from 27°N to 25°N latitude; and the TDF Southern Phase, from 26°N to 21°N latitude. We will focus on crops of the first two subzones, where summer rainfall during the growing season ranges between 293 and 471 mm (Wiseman 1980).

MAIZE

The prevailing maize *(Zea mays)* folk varieties of the TDF zone are closely related: chapalote, reventador, blando de Sonora, and onaveño. Dulcillo del noroeste, tabloncillo, and tuxpeño are also present. Reventador and chapalote are extremely ancient in the region and may have been the first maize land races to arrive in the present-day southwestern United States through prehistoric cultural diffusion. Nevertheless, the maize races of northwestern Mexico may be the least understood of all the Mexican maize races in terms of their origin and evolution.

Chapalote, one of four recognized ancient indigenous races, is thought to have derived directly from the primitive pod-pop corns (Wellhausen et al. 1952). Archaeological evidence of chapalote-like cobs from caves in New Mexico and central and eastern Mexico suggest that chapalote was once much more widespread than today (Mangelsdorf 1974), when it is known primarily from Sonora and Sinaloa. Another indication of its once wide-spread distribution is the wide range of elevations to which it is adapted; it produces ears at elevations up to 2,200 m but performs best at low elevations (100–600 m). Chapalote is a popcorn (a primitive trait) and is easily recognized by its color, which varies from light caramel to dark chocolate, and by its small cigar-shaped ears with small flinty kernels that commonly show prominent husk striations. The Mayo and Tarahumara grind toasted chapalote and use it to make *pinole*, a gruel-like drink.

Reventador, another maize race found in the TDF zone, is very closely related to chapalote. Reventador is thought to have resulted from teosinte introgression with chapalote because they share many characteristics. Both were once more widespread and overlapping in their distributions. Reventador is classified as a pre-Columbian race and is a purported parent for several maize races that are well established in the region, namely dulcillo del noroeste and tabloncillo. Reventador clearly resembles chapalote: the cobs are cigar shaped and taper slightly at both ends, and the cream- to white-colored kernels are small and rounded and typically bear pronounced striations. It is also a popcorn, as evidenced by the derivation of the name *reventador* from the Spanish verb *reventar*, which means "to burst or pop" (Wellhausen et al. 1952). Reventador is also used for pinole.

Blando de Sonora is a floury-kerneled maize found at lower elevations (100–500 m) in Sonora. It strongly resembles the flour corns of the O'odham tribes in southern Arizona and is similar to prehistoric maizes that date to A.D. 500–700 (Wellhausen et al. 1952). Blando de Sonora is ground for flour and used for making cookies.

Onaveño is classified as a popcorn and is characterized by cream to light yellow, flinty kernels. Its ears are similar in shape to those of reventador, though they are typically shorter and thicker. It is an all-purpose crop within the Guarijío/Mayo region of southern Sonora and is eaten as *elote* (corn on the cob) and used to make tamales, *atole* (a beverage), and pinole. It is also cultivated at lower, warmer elevations within the barrancas of the Sierra Madre Occidental.

Dulcillo del noroeste, a sweet corn, is adapted to low elevations in northwestern Mexico and strongly resembles reventador, with its long, slender ears that taper slightly at both ends and its small, pale yellow kernels. Cobs with burnt orange to red kernels have been collected among the Guarijío in southern Sonora. Dulcillo del noroeste is undoubtedly related to maiz dulce de Jalisco, which typically ranges in color from pale yellow to bright orange-yellow to deep red (Wellhausen et al. 1952).

Tabloncillo is a low-elevation maize found entirely on the west coast of Mexico from Sonora south to Nayarit and Jalisco (Sanchez and Goodman 1992). A subrace, tabloncillo perla, is characterized by a hard, flinty endosperm and is prevalent at lower elevations, though it has been found growing up to 1,200 m. It has been suggested that the flinty endosperm in tabloncillo perla was selected at lower elevations because it provides greater resistance to weevil predation. At higher elevations in the Sierra Madre Occidental, apachito and cristalino de Chihuahua, both of which have hard, flinty kernels, are preferred over the soft-floured maiz azul for the same reason.

The tuxpeño-like dent corns from northwestern Mexico show affinities to both tuxpeño and vandeño. The latter is the Pacific Coast equivalent of tuxpeño but reportedly matures earlier, is less vigorous, and has somewhat thicker cobs and more rows than tuxpeño (Wellhausen et al. 1952). Collections of purported tuxpeño have been made from the Mayo, Guarijío, Yaqui, and Mountain Pima regions, where few tuxpeño collections were previously recorded. Following Ortega Paczka (1985), these collections have been classified as tuxpeño norteño, though a more thorough evaluation of this material might help refine their racial affinity.

Other land races are grown in the Sierra Madre Occidental today, particularly at higher elevations. These include maiz azul, apachito, cristalino de Chihuahua, and gordo, all of which are closely related and may represent nothing more than flint (apachito and cristalino de Chihuahua) and flour (maiz azul and gordo) versions with the same genetic makeup (Sanchez and Goodman 1992).

Earlier scientific opinion that many Mexican land races exhibited recent evidence of teosinte introgression and enrichment has now been tempered by evidence from teosinte-infested fields in Nabogame, Chihuahua (Doebley and Nabhan 1989). Allozyme analyses of teosinte, maize exposed to teosinte in Nabogame, and unexposed maize from Baborigame show that few northern teosinte traits persist in maize, and vice versa. In other words, maize and teosinte growing in the same field do not become allozymically homogeneous, even though they readily hybridize. This lack of retention of introgressed alleles is presumably due to disruptive selection after initial hybridization and allows the maintenance of distinct genetic constituents for loci other than those controlling the differences in ear morphology (Doebley and Nabhan 1989).

Nevertheless, Northern Tepehuán farmers from TDF zones in nearby barrancas still bring their maize to Nabogame to have it revitalized in fields where teosinte is intentionally protected (Wilkes 1985). Reports of other teosinte populations in Tarahumara and Mountain Pima country remain unconfirmed but plausible. In summary, TDF-zone maize land races were spread prehistorically to much of the southwestern United States and are highly adapted to rapid growth following germination with the first summer rains. It remains doubtful that sporadic persistence of teosinte genes in these land races has contributed much to their genetic diversity; adaptation to a heterogeneity of agrohabitats and culinary uses appears to be more important.

SONORAN PANICGRASS

Panicgrass *(Panicum sonorum)* was once grown by tribes ranging from the Colorado River to the foothills of the Sierra Madre Occidental and beyond. Along the Colorado River, the Cocopa once cultivated panicgrass that appeared to differ from the domesticated panicgrass grown by the Guarijío, Mayo, Opata, and Lowland Pima/Nevome in the TDF zone, at least on the basis of grain size, color, and sweetness; nonstaggered germinability; fewer, more synchronous tillers; and panicle gigantism (Nabhan and de Wet 1984). Newer evidence, however, suggests that the now extinct panicgrass grown by the Cocopa may be indistinguishable from that grown in the TDF zone, though their precise relationship remains undetermined (Felger, in press). Today only a few Guarijío in the barrancas along the Sonora-Chihuahua border grow this once-widespread crop, and in a sense, the Sierra Madre Occidental has served as a refugium for its agroecological persistence (Nabhan 1985a). Already rare by 1978, Sonoran panicgrass has been further imperiled since then by the invasion of exotic buffelgrass *(Pennisetum ciliare)* into the slash-and-burn fields and by adjacent secondary vegetation in the TDF zone.

GRAIN AMARANTHS

Two distinct domesticates of grain amaranths *(Amaranthus cruentus* and *A. hypochondriacus)* have been grown in the TDF zone of the northern Sierra

Madre Occidental, but both may have in their origins or subsequent evolution genes from weedy *A. hybridus*. The latter weed is so common in Guarijío, Mayo, Lowland Pima, and Yaqui fields that introgression with the domesticated amaranths may have been widespread (Nabhan 1979).

Today grain amaranths are only occasionally found as a garden crop among the Mountain Pima and Northern Tepehuán and are found as a minor field crop in the TDF zone only among the Guarijío. Amaranth seeds were historically used for pinole, and the leaves were eaten as *quelites* (cooked greens) by the Yaqui, Mayo, and Lowland Pima. Among the Guarijío, the seeds are also used for tamales or popping. A black-seeded Mayo variety is used for atole and *esquite* (parched seeds used to make pinole).

GRAIN CHENOPODS

Although well known from central Mexico, a grain chenopod domesticate *(Chenopodium berlandieri)* has only recently been recorded from the U.S. Southwest and northwest Mexico. It may have historically grown as far north as the Pima River of central Arizona (Rea 1991) but now appears restricted to a few refugia among the Mayo, where the seed is toasted and used for pinole and atole. Pennington (1982) suggests that historically, the Eudeve may have grown a chenopod, which they called *vacue*.

BEANS

Tepary beans remain the most distinctive element of the legume crop assemblage in the northern Sierra Madre Occidental, yet common beans continue to be much more widely cultivated in the region. White teparies, pintos, and azufrado common beans dominate Indian fields in the TDF zone. Runner beans are less common at lowland elevations. One lima bean *(Phaseolus lunatus)* and one brown tepary bean have been collected from the Yaqui, but they remain poorly known farther back in the sierra.

Although 19 species of wild beans grow in the region, there have only recently been reports of introgression between wild runner beans and related domesticated land races of *P. coccineus* at higher elevations in the northern Sierra Madre Occidental (Robert Bye, pers. comm. 1997). A single notice of a wild bean *(P. salicifolius)* being cultivated among TDF-zone barranca dwellers appeared in Pennington (1969). Gary Nabhan and Barney Burns interviewed Northern Tepehuán who were living in barrancas near the Sinaloa-Durango boundary, and they confirmed this practice. No viable specimens have been collected, however.

SQUASH

The highly variable, green-striped cushaw *(Cucurbita argyrosperma)*, considered the most drought and heat tolerant of the domesticated cucurbits, is the most common TDF zone species, followed by the segualca, or big cheese pumpkin *(C. moschata)*. Although acorn squashes *(C. pepo)* grew in the region very early, they fare better at higher elevations and are seldom common in the TDF zone today.

Nabhan (1984) fully documented introgression between green-striped cushaws and conspecific feral gourds *(C. palmeri)* in Lowland Pima fields at Onavas, Sonora. Merrick and Nabhan (1984) obtained fragmentary corroborative evidence that the same field-side hybridization occurs in Mayo, Mountain Pima, and Guarijío fields. Although fig-leafed gourds *(C. ficifolia)*, chayote *(Sepium edulis)*, acorn, and blue hubbard *(C. maxima)* squashes are now found at higher elevations in Mountain Pima, Tarahumara, and Tepehuán country, they appear to be recent introductions.

BOTTLE GOURDS

Bottle gourds are highly variable in shape and size in the TDF zone. They are still widely grown and much used as containers, rattles, water drums, and canteens. However, seed morphological characteristics demonstrate limited variation relative to that seen in other parts of the world for this, the most ancient of crops (Heiser 1973). In the TDF zone, the Mayo and canyon Tarahumara appear to cultivate bottle gourds with the greatest diversity in shape and size, though this perceived trend could be an artifact of NS/S collection trips.

CHILE PEPPERS

Virtually every tribe in the northern Sierra Madre Occidental grows chiles as a garden or field crop, but it is unclear which folk varieties are the most ancient. Much of the saved seed produces plants that show characteristics of accidental hybridization with improved varieties derived from plant-breeding programs in New Mexico and California. In the TDF zone, however, chile land races also display evidence of inadvertent introgression with wild chiltepines *(Capsicum annuum* var. *aviculare)* in gardens at Onavas, Sonora, where Lowland Pima descendants recognize them by their extreme piquancy and diminished size. The widespread distribution of chiltepines in the TDF suggests that this same genetic exchange also occurred within Mayo, Guarijío, and Yaqui fields.

TOBACCOS

Two tobaccos persist in most Indian villages in the TDF zone, but the strongest ceremonial traditions are linked to makuchi *(Nicotiana rustica)*, which includes enough nor-nicotine that some observers have suggested it is mildly hallucinogenic. The tobacco more commonly grown in Mexico by mestizos, *Nicotiana tabacum*, is now used in the absence of makuchi in *pascola* and deer dancer ceremonies among the Guarijío, Mayo, Yaqui, and Tarahumara.

CONIVARI

Also known as *chia grande* among mestizos, this pseudograin *(Hyptis suaveolens)* is used much the same way as *Salvia* species, which are known as *chia*. The cultivated plant may be an incipient domesticate (Gentry 1963), which is grown in dooryard gardens of the Guarijío and Mayo a few hundred kilometers north of where it grows as a common weed in the TDF zone.

AGAVES

The progenitor of tequila, *Agave angustifolia*, appears to have been independently cultivated in the northern Sierra Madre Occidental. However, it has not been as fully domesticated or as rigorously selected for shorter maturation time as Bye et. al. (1975:95) reported for the western Tarahumara:

> Found in the lower parts of the arid sub-tropical barrancas in short thorn forest, *Agave pacifica [sic]* produces sucker shoots which are transplanted near dwellings in the lower oak *(Quercus albocincta)* zone of the middle barranca zone. Considered the most delicious and difficult to collect in the field, this species is often grown near houses. It also became popular with the Spaniards and Mexicans who settled in the deep canyons and later transplanted and cultivated it. The hearts of the cultivated plants are larger than those of the wild plants.

The Lowland Pima cultivate this species for *bacanora*, a bootleg tequila, at *rancherías* (groups of huts) near Onavas, Sonora, as do mestizos of Mayo descent who live between Alamos and San Bernardo, Sonora. It is overharvested from the wild in many parts of Sonora and Chihuahua; its populations are reported to be too small to harvest near San Francisco de las Pimas.

OTHER NATIVE CROPS

A number of other native crops may have been grown in indigenous gardens and fields within the region. For example, saiya or temaqui *(Amoreuxia palmatifida)* has been grown in Mayo dooryard gardens in San Bernardo and fits the description of an early historical root crop grown by the Nevome and Eudeve in Sonora. Around Masiaca, Sonora, *A. gonzalezii* is the more common species, although both grow in the area. The Mayo consider the two species distinct enough to have separate names, referring to *A. gonzalezii* as *saya mome.*

The incipient domesticates of onions, mustards, and solanaceous quelites found above the TDF zone in the northern Sierra Madre Occidental should be sought within the barrancas. Other New World domesticates — from sunflowers to calabash trees — grow in fields and gardens not far from the TDF zone in the northern Sierra Madre Occidental, and their products are occasionally used by indigenous peoples in this zone. It is likely that ethnographers and crop geographers have overlooked their presence in the region.

Factors That Promote Crop Diversity

As stated earlier, it is difficult if not impossible to quantitatively measure the relative importance of factors that contribute to the astounding diversity of native crop species and folk varieties in the TDF zone of the northern Sierra Madre Occidental. The following discussion highlights the relative importance of physical, biological, agroecological, and cultural factors that shape and maintain crop diversity in the region.

PHYSICAL FACTORS

The ruggedness of terrain in the northern Sierra Madre Occidental, the relative isolation of each valley and barranca, and the difficulty of travel between certain areas have no doubt allowed for microevolution of certain crop plants once they were brought into cultivation. In addition, the steepness of precipitation gradients and the differences in temperature may force farmers who live just a few kilometers from each other (but at different elevations) to select entirely different crop species or varieties adapted to different temperature regimes or summer rainfall intensities. For example, the members of one Tarahumara family who live at two different elevations within the barranca system of the Sierra Madre Occidental manage to cultivate at least

10 maize races between them. Sweet corn (dulcillo del noroeste), popcorn, harinoso de ocho, tabloncillo, and tuxpeño norteño are grown at a lower elevation (1,620 m) on their farm in Cerocahui, whereas none of these crops are grown at 2,190 m on their farm in Cusarare. In contrast, apachito, cristalino de Chihuahua, and maiz azul are grown at both elevations.

Soils may also affect crop survival and production. In a Mountain Pima ranchería near Maicoba, Sonora, we observed that common beans were dying of nutrient deficiencies on a volcanic ash slope, but wild beans were quite healthy on the same soil. Edaphic extremes may favor microevolution of adapted crop genotypes.

BIOLOGICAL FACTORS

The TDF zone is rich in wild congeneric relatives of domesticated crops, and some may exchange genes with their cultivated counterparts through inadvertent or managed introgressive hybridization. The following crops may freely exchange genes with wild relatives that are native to the northern Sierra Madre Occidental: maize (at Nabogame and perhaps other sites), Sonoran panicgrass, grain amaranths, tepary and scarlet runner beans, green-striped cushaw squash, chile peppers, and agaves. This gene exchange may help these crops to cope better with diseases, pests, or physiological constraints specific to the bioclimatic zone.

Gene exchange between different races of the same crop, particularly maize, may have been equally important in contributing to crop diversity within the region. Although the exact origin and evolutionary history of individual maize races remain vague, groupings of particular races consistently appear to be related (Doebley et al. 1985; Sanchez and Goodman 1992). Chapalote and reventador appear to be very closely related and may have been important in the formation of other races in northwestern Mexico. It has been suggested that blando de Sonora may have resulted from introgression of chapalote and reventador into harinoso de ocho (an eight-rowed, low-elevation flour maize now mostly replaced by one of its other derivatives, tabloncillo). In turn, onaveño may be derived from introgression of reventador into blando de Sonora (Sanchez and Goodman 1992). Reventador has been postulated to be one of the parents of both dulcillo del noroeste and tabloncillo (Mangelsdorf 1974; Wellhausen et al. 1952).

Thus, a great deal of gene exchange has occurred, both between wild and domesticated crops and between different races of the same crop. The great diversity of maize races in Mexico suggests that gene exchange between races may have been more important than that between domesticates and their wild relatives, particularly as the native habitats that nurtured the crop precursors became scarce.

AGROECOLOGICAL FACTORS

Farmers, gardeners, and orchard growers have created an astonishing variety of agrohabitats within the TDF zone: floodplain and flood recession fields; river diversion, ditch- or border-irrigated, floodplain fields; *ak-chin* or *de temporal* floodwater spreading fields on alluvial fans; rain-fed, slope-side, slash-and-burn fields; well-supplemented dooryard gardens; irrigated tree-well plantings of herbaceous crops within orchards on mesa tops and in valleys; and ridge-top, hillside terraces or hedgerow plantings of agaves. These agrohabitats favor certain crop mixtures and spacings over others.

Some TDF-zone farmers and gardeners routinely intercrop their native domesticates rather than grow them in monocultural stands, potentially maximizing use of their limited land resources. Plant traits, such as growth habit, days to maturity, and shade tolerance, may dictate which crops may be intercropped (Nelson and Robichaux 1997). Both intra- and interspecific competition or facilitation (depending on the cropping system) may affect crop selection within this region.

The effective use of intercropping depends primarily on minimizing competition between the component crops in time and/or space (Nelson and Robichaux 1997). A 92-year-old farmer in northern Sonora related how his father, who was originally from the highlands of Oaxaca, taught him to plant maize and bean simultaneously in the same hole with a planting stick. They knew that this method worked for only certain combinations of maize and bean: some beans were better planted with maize; others were not. The latter type of bean needed to be planted when the maize was of a certain height or maturity. For maize/bean intercrops in the northern Sierra Madre Occidental, the tendency is to interplant beans among maize that has been growing for several months.

In both cases competition for plant resources, such as water, light, and nutrients, may be minimized through temporal differences in the peak resource demands of the individual crops, though this may be achieved through drastically different means in each case. In simultaneous planting, temporal differences in peak resource demand may result from the use of folk varieties with intrinsically different maturity times. In staggered planting, offsetting peak resource demands may result from simply giving the

Table 5.2. Native Speakers and Inhabitants of the Northern Sierra Madre Occidental

Indigenous Group	Total Maximum Population	Current Speakers	States in Residence
Tarahumara	66,000	65,160	Chihuahua
Mayo	56,387	45,195	Sonora, Sinaloa
Yaqui	25,000	12,011	Chihuahua, Sonora
Northern Tepehuán	14,900	2,959	Chihuahua, Sinaloa
Guarijío	5,000	3,000	Chihuahua, Sonora
Mountain Pima	2,006	124	Chihuahua, Sonora
Lowland Pima	553	4	Sonora
Opata	< 300	?	Sonora
Total	170,146	128,453	

Source: Jane Hills and Emilia Estrada, pers. comm. 1997.

maize a head start. Use of different cropping systems, and hence the need for specific crop types, may also have contributed to overall crop diversity within the region.

CULTURAL FACTORS

The 180,000 km^2 of the northern Sierra Madre Occidental is currently home to more than 170,000 individuals of indigenous descent (table 5.2). At least 128,000 of these speak eight native languages (Fowler 1996; Gary Nabhan, unpubl. data). Each of these languages encodes biological, agronomic, and culinary information about crops that may drive native crop domestication, selection, hybridization, or isolation in dramatic ways (Nabhan and Rea 1987). The indigenous names for certain crops or varieties (appendix 5.1) may suggest routes or relative antiquities for crop diffusion from one tribe to another, but such clues should be interpreted in light of other genetic, historical, and cultural geographic data. In addition, it is important to remember that the current level of cultural diversity in the region is but a dim shadow of what it once was. At the time of the Pérez de Ribas (1645) *entradas* (entries or assaults) into the TDF zone, many additional cultural communities were recognized in the region. They included the Eudeve, Cinaloa, Joba, Janos, Tubares, and Chinipas communities, among others. How the TDF may have affected or been affected by the agricultural practices of these groups is virtually unknown. These groups, however, undoubtedly influenced prehistoric crop diversity. Perhaps some remnants of their land races are still found among the surviving groups.

Genetic studies on the evolution of and relationships among different races of a domesticated crop may also shed light on its historical or current distribution. Research on maize land races suggested three potential migration paths for maize germ plasm in North America: a route from Guatemala to Sonora along the Pacific coast; a route from central Mexico to Chihuahua; and a more recent route along the east coast of Mexico from southern Veracruz to Tamaulipas (McClintock 1978). The exact manner in which maize or any other domesticated crop migrated from one place to another may never be known, though it was undoubtedly facilitated through cultural migrations.

Perhaps more than any other factors, differences in cultural practices, preferences, and taboos have contributed to the dazzling display of crop types, seed colors and sizes, flavors, pungencies, fruit shapes, etc. Domesticated crops cannot survive without intervention from humans and have thus evolved as a direct consequence of the interaction between human and natural selection pressures. It can at least be argued that the very diversity of cultures and the trade between them in the northern Sierra Madre Occidental have not only fostered crop diversity but also helped it to persist more than it has in adjacent, more culturally homogeneous regions.

Conclusions

The continued survival of native crop diversity in the TDF zone is dependent upon slowing or arresting the processes that might otherwise disrupt the factors that have promoted and maintained this diversity. For example, the geographical isolation of farming valleys can be broken by the encroachment of new roads into the region, such as those used for logging or mining. Setting up roadless biosphere reserves, however, will help to maintain the isolation of some areas. If fields become too large or if hedgerows and patches of wild crop relatives are plowed out of existence, weed/crop introgession will decline. Traditional agrohabitats need protection as much as wilderness areas do.

Who will tend those traditional agrohabitats? The need for supporting and protecting the lifestyles of traditional agriculturalists is just as compelling as the need for wilderness protection. These farmers need access to loans and markets. Agrarian and banking laws might be restructured so that small landholders are not forced to sell their lands or switch to capital-intensive industrial agriculture. Equally destructive is the full-scale arrival of the drug industry and the lack of choices it presents. Can a Mayo farmer choose between the meager subsistence life he has always known and a life in which he earns more in one year than he would in a lifetime simply by guarding his milpa-turned-opium field? For many, good intentions or not, this is simply not a choice. To whom will he pass on the wealth of accumulated agricultural knowledge gathered from generation to generation of farmers? In what language will this transfer of information take place, and will all the subtleties of one language be accurately captured in another? Cultural and linguistic preservation is thus essential to maintaining the agricultural diversity of the area. In this sense, the future of crop diversity is not as dependent upon an *ex situ* gene-banking program as it is upon the survival of *in situ* indigenous farming communities.

ACKNOWLEDGMENTS

This paper is dedicated to the many native farmers who have shared their agricultural knowledge and seed samples with us over the years. We would also like to thank our colleagues—Robert Bye, Howard Scott Gentry, Campbell Pennington, Peter Bretting, Felipe Molina, Edmund Faubert, Amadeo Rea, Linda Parker, Jose Muruaga Martinez, Martha Burgess, Eric Powell, Richard Felger, Eric Mellink, Joseph Laferriere, Jane Hill, Catherine Fowler, John Doebley, Garrison Wilkes, Steve Tanksley, and Fernando Loiza-Figeroa—for field and laboratory collaboration. Some of this work was made possible by National Science Foundation grants to Gary Nabhan, Amadeo Rea, and Richard Felger and by several private foundation grants to Mahina Drees, Barney Burns, Gary Nabhan, Suzanne Nelson, and Robert Robichaux.

REFERENCES

Alcorn, J. B. 1984. *Huastec Mayan ethnobotany*. Austin: University of Texas Press.

Bye, R. A., Jr. 1995. Phytogeography of the Sierra Madre Occidental pine-oak woodlands and its contribution to Mexico's biodiversity. In *Madrean woodlands*, eds. P. Ffolliott and L. Bain, 19–27. Fort Collins, Colo.: U.S. Forest Service.

Bye, R. A., Jr., D. Burgess, and A. Mares-Trias. 1975. Ethnobotany of the western Tarahumara of Chihuahua, Mexico: I. Notes on the genus *Agave*. *Harvard University Botanical Museum Leaflets* 24:85–112.

Doebley, J., M. M. Goodman, and C. W. Stuber. 1985. Isozyme variation in races of maize from Mexico. *American Journal of Botany* 72:629–639.

Appendix 5.1. Indigenous Names for Native Wild and Domesticated Crops of the Northern Sierra Madre Occidental

Crop Species/Race	Northern Tepehuán	Tarahumara	Guarijío	Yaqui & Mayo	Lowland Pima	Mountain Pima	Opata & Eudeve
Agave angustifolia			temechi, chahuiqui	wah kuu'u	su'ut	ma'i	vitzo
Amaranthus cruentus							
Amaranthus hypochondriacus	giagi		guegui	wee'e	guiguida		gueró
Brassica campestris		mocoasali			okiti	ckiti	
Capsicum annuum		kori sitakame		ko'okoi	ko'okor	kokol	uruc
Chenopodium berlandieri				bwaarom	guiguida (?)		vacue
Cucurbita argyrosperma	babakilimai			kuta kama	ha'al, baviri	iim	
Cucurbita ficifolia	pipintimai, xiotai						
Cucurbita moschata	mumuiala			segualca	xa'askas haal	iim	cama, babiris, sosoc, bara
Cucurbita pepo	iimai		ha'la we		tongidat ha'as	iim vavuli	vavora (?)
Helianthus annuus				taa'atavitchu			

			conivari, cham	conivari			
Hyptis suaveolens							
Jaltomata procumbens	oidigana						
Lagenaria siceraria	vakoi, vacamai,	makuchi		visa'e	haavui, vako	vak, havo	ari
Nicotiana rustica	makuchi		makuche	makucha	vibbac	simblon viv	
Nicotiana tabacum	vivai		wipa	viiva	o'atk	viv	vivat
Panicum sonorum			sagüi		gugurhsabi	sabi	
Phaseolus acutifolius			serowi	se'elaim, heseim	bavi		tepar
Phaseolus coccineus	tukamuli	tekomari			vi posol		
Phaseolus salicifolius	tupuuli						
Phaseolus vulgaris	bavi	muni	muni	muuni	bav		mun
Zea mays	uunui			batchi	hun		sunut

Sources: Indigenous crop names follow Fowler (1996), Gentry (1963), Molina and Shaul (1993), and Pennington (1963, 1969, 1970, 1982, 1989).

Doebley, J., and G. P. Nabhan. 1989. Further evidence regarding gene flow between maize and teosinte. *Maize Genetics Cooperative Newsletter* 63:107–108.

Felger, R. S. In press. *Flora of the Gran Desierto and Río Colorado of Northwestern Mexico*. Tucson: University of Arizona Press.

Felger, R. S., G. P. Nabhan, and R. A. Bye Jr. 1997. Apachian/Madrean region of south-western North America. In *Centres of plant diversity*, eds. S. D. Davis, V. H. Heywood, O. Herrera-MacBryde, J. Villa-Lobos, and A. C. Hamilton, 172–180. Oxford: Oxford University Press.

Fowler, C. 1996. Linguistic diversity in the greater Southwest. Paper presented at the symposium, Losing Languages, Species, and Stories, April 1996, Arizona-Sonora Desert Museum, Tucson.

Gentry, H. S. 1963. The Warihío Indians of Sonora-Chihuahua: An ethnographic survey. Anthropological Papers No. 65. *Bureau of American Ethnology Bulletin* 186:61–144.

Heiser, C. 1973. Variation in the bottle gourd. In *Tropical forest ecosystems in Africa*, eds. B. Meggers, E. Ayensu, and N. D. Duckworth, 121–128. Washington, D.C.: Smithsonian Institution Press.

Mangelsdorf, P. C. 1974. *Corn: Its origin, evolution, and improvement*. Cambridge: Harvard University Press.

McClintock, B. 1978. Significance of chromosome constitutions in tracing the origin and migration of races of maize in the Americas. In *Maize breeding and genetics*, ed. D. B. Walden, 159–184. New York: Wiley.

Merrick, L. C., and G. P. Nabhan. 1984. Natural hybridization of wild *Cucurbita sororia* group and domesticated *Cucurbita mixta* in southern Sonora, Mexico. *Cucurbita Genetics Cooperative Newsletter* 7:73–75.

Molina, F. S., and D. L. Shaul. 1993. *A concise Yoeme and English dictionary*. Tucson: Tucson Unified School District.

Nabhan, G. P. 1979. Amaranth cultivation in the U.S. Southwest and northwest Mexico. *Rodale Press Amaranth Conference Proceedings* 2:129–133.

———. 1984. Evidence of gene flow between cultivated *Cucurbita mixta* and a field edge population of wild *Cucurbita* at Onavas, Sonora. *Cucurbita Genetics Cooperative Newsletter* 7:76–77.

———. 1985a. *Gathering the desert*. Tucson: University of Arizona Press.

———. 1985b. Native crop diversity in Aridoamerica: Conservation of regional gene pools. *Economic Botany* 39:387–399.

———. 1990. *Wild* Phaseolus *ecogeography in the Sierra Madre Occidental, Mexico*. Systematic and Ecogeographic Studies on Crop Genepools No. 5. Rome: International Board for Plant Genetic Resources.

Nabhan, G. P., and J.M.J. de Wet. 1984. *Panicum sonorum* in Sonoran Desert agriculture. *Economic Botany* 38:65–82.

Nabhan, G. P., and R. S. Felger. 1978. Teparies in southwestern North America: A biogeographical and ethnohistorical study of *Phaseolus acutifolius*. *Economic Botany* 32:2–19.

Nabhan, G. P., and A. M. Rea. 1987. Plant domestication and folk biological change: The Upper Pima/devil's claw example. *American Anthropologist* 89:57–73.

Nelson, S. C., and R. H. Robichaux. 1997. Identifying plant architectural traits associated with yield under intercropping: Implications of genotype-cropping system interactions. *Plant Breeding* 116:163–170.

Ortega Paczka, R. A. 1985. Variedades y razas Mexicanas de maiz y su evaluación en cruzamientos con lineas de clima templado como material de partida para fitomejoramiento. Abbreviated Spanish translation of Ph.D. dissertation, N. I. Vavilov National Institute of Plants, Leningrad.

Pennington, C. W. 1963. *The Tarahumara of Mexico: Their environment and material culture*. Salt Lake City: University of Utah Press.

———. 1969. *The Tepehuán of Chihuahua: Their material culture*. Salt Lake City: University of Utah Press.

———. 1970. A vocabulary made at Maicoba, Sonora, among the Pima Bajo (1968, 1970). Unpublished manuscript, Southern Illinois University, Carbondale.

———. 1982. *La cultura de los Eudeve del Noroeste de México*. Noroeste de México No. 6. Hermosillo: Centro Regional del Noroeste, INAH-SEP.

———. 1989. *The Pima Bajo of central Sonora, Mexico. Volume I: The material culture*. Salt Lake City: University of Utah Press.

Pérez de Ribas, A. 1645. *Historia de los triumphos de nuestra Santa Fe entre gentes las más bárbaras y fieras del Nuevo Orbe*. Madrid: A. de Paredes. Translated by D. T. Reff, M. Ahern, and R. K. Danford under the title *History of the triumphs of our Holy Faith amongst the most barbarous and fierce peoples of the New World* (Tucson: University of Arizona Press, 1999).

Rea, A. M. 1991. Gila River dietary reconstruction. *Arid Lands Newsletter* 31:3–10.

Sanchez, J. J., and M. M. Goodman. 1992. Relationships among the Mexican races of maize. *Economic Botany* 46:72–85.

Wellhausen, E. J., L. M. Roberts, and X. E. Hernandez. 1952. *Races of maize in Mexico*. Cambridge: Harvard University Press.

Wilkes, W. G. 1985. Teosinte: The closest relative of maize revisited. *Maydica* 30:209–223.

Wiseman, F. M. 1980. The edge of the tropics: The transition from tropical to subtropical ecosystems in Sonora, Mexico. *Geoscience and Man* 21:141–156.

6

AMPHIBIANS AND REPTILES OF
THE SIERRA DE ALAMOS

Cecil R. Schwalbe and Charles H. Lowe

The remarkable biotic diversity in southern Sonora has lured biologists and nature lovers for decades. This allure may be heightened near the northern limits of the tropical deciduous forest (TDF), where contrasts between lush tropical communities and harsher, more xeric thornscrub and desertscrub landscapes at lower elevations and more northerly latitudes are so dramatic.

The area around Alamos, Sonora, is rapidly being discovered by increasing numbers of herpetologists seeking some of the most interesting amphibians and reptiles in northern Mexico. Long known for fantastic birding, spectacular vegetation and landscapes, and a rich human history, this area is now being targeted by those coming to see, photograph, and sometimes take an incredibly rich herpetofauna. The availability of inexpensive to posh tourist accommodations makes a visit to the area even more appealing to those not wishing to rough it. One can pitch a tent or plug in a trailer at any of several trailer parks or stay at any of numerous hotels ranging from minimal lodging to the lap of luxury, all in seventeenth-century-style architecture.

Hilton's (1947) *Sonora Sketch Book* provides enjoyable glimpses of a slower-paced life, including herpetofaunal collecting, in southern Sonora a half century ago. One cannot wander with the same abandon in the 1990s; the possibility of accidentally stumbling upon a field of marijuana while looking for lizards gives one pause for thought. *The Secret Forest* (Bowden et al. 1993) focused attention on threats to the TDF, threats that are expanded upon elsewhere in the present volume.

This chapter focuses on the rich variety of amphibian and reptile species in the Sierra de Alamos area (appendices 6.1–6.3). However, because of the interest in the diverse biota of this part of northern Mexico, we have also compiled from the literature a list of mammal species that live in the region (appendix 6.4). Although we are fascinated by some of the mammal species inhabiting southern Sonora, most of our observations of the mammals are purely anecdotal. We have not conducted systematic searches or surveys for mammals as we have for amphibians and reptiles. We know of no extensive recent mammal inventories in the Alamos region, although several biologists have netted bats in the area. In Mexico, tropical areas contribute the largest number of species to the country's high diversity of mammals: 449 species composed of mostly rodents and bats (Fa and Morales 1993). The contribution of tropical species to the diversity of mammals in the Alamos area is also significant (Burt 1938; Caire 1978; Goldman 1951). Thorough surveys of the area for mammal species would be a valuable addition to our knowledge of the region.

As with many herpetofauna species, several mammal species reach their extreme northern or northwestern distributions in TDF in the Alamos area. Notable examples include the grayish mouse-opossum *(Marmosa canescens)*; several species of bats, including the vampire bat *(Desmodus rotundus)*; and the nine-banded armadillo *(Dasypus novemcinctus)*. Other species, such as the desert shrew *(Notiosorex crawfordi)*, Botta's pocket gopher *(Thomomys bottae)*, and the California leaf-nosed bat *(Macrotus californicus)*, reach the southern limits of their continental ranges in southern Sonora, Chihuahua, and northern Sinaloa.

Sources of Information

Our involvement with this diverse herpetofauna extends back to one of the earliest intensive collecting expeditions to the area (Bogert and Oliver 1945), a trip attended by one of us (Charles Lowe) as a field assistant. Most of the knowledge of the biota of this area has accrued since the 1930s. Bogert and Oliver (1945) presented the first detailed look at the TDF herpetofauna in southeastern Sonora. Smith and Taylor's (1945, 1948, 1950, 1966) checklists and keys summarized what was known about the identity and distribution of amphibians and reptiles in Mexico but did not provide much new information on habitat affinities or ecology. Flores-Villela (1993a,b) was the first to attempt to characterize the habitats, distribution, and endemism of the entire Mexican herpetofauna. The information presented in this chapter is based upon relatively few published accounts, which are mostly distributional, and upon many observations by us and our associates over the past 50 years, especially during the last two decades. Detailed studies on the ecology and demography of the Sonoran herpetofauna await completion.

Herpetofaunal Diversity and Habitats

For this chapter we are defining the "Alamos area" as the mountain mass of the Sierra de Alamos plus the surrounding areas that extend to the Río Mayo to the north, the Río Cuchujaqui to the east and south, and the western edge of the Sierra de Alamos to the west and south near Masiaca (see map in preface). This area, approximately 30 × 40 km in size, ranges from about 100 to 1,700 m in elevation and comprises less than 1% of the area of the state of Sonora. Within this relatively tiny area, 79 species of amphibians and reptiles have been found (appendices 6.1 and 6.2)—49% of the 160 species that live in the entire state of Sonora!

To give some additional perspective on how diverse the herpetofauna is in the Alamos area, we will compare some of these numbers to those in Arizona (table 6.1). Arizona has a well-earned reputation as one of the "herping hot spots" in the United States. People come from all over the world to search for and admire the native Arizona herpetofauna. Although it includes only about 3% of the combined area of the 48 contiguous states, Arizona contains more than one-fourth (122) of the 475 species of amphibians and reptiles found in the United States. In comparison, Mexico supports almost 1,000 species (995) of amphibians and reptiles, 16% of which live in Sonora.

Sonora has about 1.3 times as many species of amphibians and reptiles as Arizona in about 0.63 times the area. Sonora, therefore, has about twice the species density (number of species per unit area) of Arizona (table 6.1). Mexico has 10 times the herpetofaunal species density of the United States, another demonstration of the incredible natural riches of the country. The Alamos area is even more diverse, with a herpetofaunal species density (65.8 species/1,000 km^2) that is 130 times that of the United States and 16 times that of Arizona! Much of this species richness is due to the presence of amphibians and reptiles that reach the northern limits of their species ranges in the TDF of southern Sonora.

We have been very restrictive in estimating species richness for the Alamos area by requiring that all species listed in appendices 6.1 and 6.2 be confirmed—by specimen, photograph, or observation—within the Alamos area. This results in a very conservative estimate of species richness for two main reasons. First, most of the Sierra de Alamos has been poorly sampled because access to higher elevations is limited. We do have many records of species observed on the Alamos-to-Navojoa road and on the many dirt roads at the base of the sierra, but higher elevations have been surveyed cursorily.

Table 6.1. Species Richness of Amphibians and Reptiles

REGION	NO. SPECIES OF AMPHIBIANS	NO. SPECIES OF REPTILES	TOTAL NO. SPECIES	AREA (km^2)	SPECIES DENSITY (# spp./1,000 km^2)
Alamos area[a]	15	64	79	1,200	65.8
Sonora	34	126	160	185,430	8.6
Arizona	25	97	122	294,000	4.1
Mexico	290	705	995	1,972,547	5.0
United States	193	282	475	9,166,350	0.5

Sources: Alamos area: Bogert and Oliver (1945), Flores-Villela (1993a), Heringhi (1969), Smith and Taylor (1966), and observations and collections by the authors; Sonora and Arizona: observations and collections by the authors; Mexico: Flores Villela (1993a); United States: Collins (1997).
[a] Includes the main mountain mass of the Sierra de Alamos as well as the surrounding areas that extend to the Río Mayo to the north, the Río Cuchujaqui to the east and south, and the western outlyers of the Sierra de Alamos to the west and south near Masiaca.

Second, the Sierra de Alamos does not have all of the habitats found in other mountain ranges in the region. In particular, it contains little perennial stream habitat, especially habitat with bedrock plunge pools, which in nearby ranges supports such species as the Tarahumara frog *(Rana tarahumarae)* and the Tarahumara salamander *(Ambystoma rosaceum)*.

Our assignment of species to certain habitats is based, somewhat subjectively, on our observations of those species over many years and on discussions with other herpetologists. We distinguish thornscrub, TDF, oak woodland, and pine-oak forest vegetation communities in the Alamos area, with distinctive wetland habitats provided by small streams, rivers, ponds, springs, and wet mines or caves. Species such as the filetail ground snake *(Sonora aemula)* and the western Mexican whiptail *(Cnemidophorus costatus)* show a distinctive preference for open and secondary growth areas. Many species, such as the Sonoran whipsnake *(Masticophis bilineatus)*, gopher snake *(Pituophis melanoleucus)*, and Clark's spiny lizard *(Sceloporus clarkii)*, act as habitat generalists in the Alamos area, living in many of the habitats available in the region. Other species are more predictably found, having much narrower habitat requirements: for example, barking frogs *(Eleutherodactylus augusti)* that most often live in wet mines or caves and Mexican leaf frogs *(Pachymedusa dacnicolor)* that breed in ponds that usually contain emergent or overhanging vegetation.

THORNSCRUB

Thornscrub grows at the lowest elevations of the upland plant communities in the region. Many of the species that we in Arizona often consider as "desert" animals actually inhabit thornscrub habitats. The paloverde/saguaro community, known as the Arizona Upland subdivision of the Sonoran Desert, is actually a depauperate thornscrub (Turner and Brown 1982). Many of the animals and plants that thrive in that biome in Arizona and northern Sonora fare quite well throughout the thornscrub proper and even up into TDF in southern Sonora and northern Sinaloa. The Sonoran Desert toad (also called the Colorado River toad, *Bufo alvarius*), regal horned lizard *(Phrynosoma solare)*, desert tortoise *(Gopherus agassizii)*, and lowland burrowing treefrog *(Pternohyla fodiens)* are classic examples.

TROPICAL DECIDUOUS FOREST

The thornscrub that is present at lower elevations within the Alamos area gives way to TDF as the land rises. This transition from thornscrub to forest is defined as the point at which the columnar cacti (hechos and organpipes) no longer extend above the tree canopy. Plant species composition often does not differ significantly across this transition zone. As tree canopy height increases above the tops of the columnar cacti, so does the number of species diagnostic for TDF.

The importance of the TDF to the diversity of biota in the region cannot be overemphasized. Many plant and animal species range from Arizona Upland desertscrub in the north to TDF in southern Sonora. Some species, such as the zebratail lizard, extend even from creosotebush flats in the Lower Colorado River Valley subdivision of the Sonoran Desert in Arizona into TDF near Alamos. However, several species can be used to define that biome.

Plants, such as amapas *(Tabebuia chrysantha, T. impetiginosa)*, cuajilotes *(Pseudobombax palmeri)*, and rattail orchids *(Oncidium cebolleta)*; and animals, such as black-throated magpie jays, military macaws, vampire bats, mouse opossums, pichecuates, parrot snakes, clouded anoles, tropical tree lizards, Mexican leaf frogs, and Mexican tree frogs, all send the message that one is in TDF. Amphibian and reptile species that live in TDF make an impressive list (appendix 6.3). Note that this list (80 species) differs slightly from that of the Alamos area (79 species; appendices 6.1 and 6.2), mainly for two reasons: (1) several species that have been observed in the Alamos area (e.g., the desert iguana *[Dipsosaurus dorsalis]* and desert spiny lizard *[Sceloporus magister]*) were never observed in TDF per se; and (2) several species found in TDF elsewhere apparently do not inhabit the Alamos area (e.g., the Tarahumara salamander, Tarahumara frog, cascade frog *[Rana pustulosa]*, and rock horned lizard *[Phrynosoma ditmarsi]*).

Appendix 6.3 is lengthy because it includes all species that live with some regularity in TDF, not just the species that are restricted to that biome. As we have indicated, it is difficult to assign most vertebrates to a single habitat or plant community. Most can survive and reproduce in several biomes. Research on packrat middens, for example, indicates that much of the present-day Sonoran Desert herpetofauna in the Tucson area was living in juniper woodlands at the same locality just a few thousand years ago (Van Devender et al. 1991).

Our collecting efforts in southern Sonora have definitely been seasonally biased. We have spent much more time searching for amphibians and reptiles from July through August than at other times of the year. We have therefore learned much about the distributions and activities of many species during the warm wet season and less about what the animals do during the rest of the year. Most of the terrestrial anurans become active for only a few months, beginning with the onset of the summer rainy season in late June or early July. However, we have found that many of the anuran species that are primarily active in TDF habitats (e.g., Mexican leaf frog, Mexican tree frog, and Sinaloa toad) will respond to fall (September–November) rainstorms, becoming briefly surface active. In fact, with the exception of a handful of species, mostly the rarer or more secretive ones (sheep frog *[Hypopachus variolosus]*, lizard eater *[Dryadophis cliftoni]*, false ficimia *[Pseudoficimia frontalis]*, and western blind snake *[Leptotyphlops humilis]*), the entire herpetofauna of the area was observed during both summer and fall. Our winter and spring trips, as well as our observations, were fewer.

Escorpiónes or Mexican beaded lizards *(Heloderma horridum)* are quite common in the TDF near Alamos. During the summer rainy season, we reliably found beaded lizards crossing roads at the base of the Sierra de Alamos in late afternoon, an hour or so before sundown. In August 1986, we saw three adult beaded lizards within 3 m of each other where the road crossed a wash about 1.5 km west of the Arroyo El Mentidero, a major watercourse that feeds the Río Cuchujaqui south of Alamos.

Near Alamos we have found adult barking frogs active from early July until mid-September in wet, rocky habitats (often wet mines and caves) in TDF or oak woodland. During summer rainstorms, they are occasionally found under rocks or are surface active, even on the road. Most often observed walking or hopping slowly in rocky habitats, barking frogs sometimes make prodigious leaps of more than 1 m.

OAK WOODLANDS

Several species of oaks combine to create different oak woodlands as one travels up the Sierra de Alamos. The relatively xeric oak woodland at lower elevations is quite distinctive from the more mesic oak communities higher up the mountain. Reptiles found in the oak woodlands include Clark's spiny lizard, Madrean alligator lizard (*Elgaria kingii*), little-eared skink (*Eumeces parviauriculatus*), and gopher snake. We also found a blackneck garter snake (*Thamnophis cyrtopsis*) on a rock pile under the oaks, far (> 0.8 km) from any wetland or riparian habitats.

Amphibians included the canyon treefrog (*Hyla arenicolor*) and barking frog. Except for one occasion, young juvenile barking frogs, with their distinctive white dorsal "sash," have been seen only singly. One afternoon in mid-September 1992, we found five juvenile barking frogs while we were hiking through oak woodland up the trail toward Aduana Peak. The frogs were active on the surface about 30 minutes before a downpour that lasted 8 hours. Interestingly, the young frogs were not near distinctive rock outcrops, as were almost all of the adult frogs that we observed.

PINE-OAK FOREST

Near the top of the mountain range, there are patches of pine-oak forest, although oaks still dominate at most sites. We found few amphibians or reptiles in our limited searches of pine-oak forest in the Sierra de Alamos. Based on observations elsewhere in Sonora, we expect eventually to find little-eared skinks, gopher snakes, and canyon tree frogs there.

On a foot search of Aduana Peak in September 1992, Bruce Palmer found a Mexican beaded lizard pawing through a fresh cow pie at the base of a pine tree at 1,400 m. This was the only beaded lizard we have found in pine-oak forest. When released after being weighed, measured, and photographed, the escorpión "escaped" from us by climbing more than 5 m up an oak tree, exhibiting its considerable arboreal talents.

OTHER HABITATS

Several distinctive wetland habitats are found in the area. The riverine habitats of the Ríos Cuchujaqui and Mayo support a herpetofauna that is different from that of smaller streams in the area. At the Río Cuchujaqui, one is much more likely to find sliders *(Trachemys scripta)*, indigo snakes *(Drymarchon corais)*, Neotropical whipsnakes *(Masticophis mentovarius)*, Nelson's spiny lizards *(Sceloporus nelsoni)*, and West Coast garter snakes *(Thamnophis valida)* than along smaller streams. The indigo snake near Alamos is most impressive. Most of our indigo snake observations were of large individuals (longer than 160 cm) foraging during the day along the rivers and major washes. We captured at least three snakes that were longer than 2 m. Several times we were alerted to the presence of an indigo snake by the release call of a leopard frog or spadefoot that the snake was swallowing. We were seldom successful in catching indigo snakes when we drove the roads looking for them. The snakes are very alert to moving vehicles and almost always escape into cover before one can stop and catch them. We had much better success catching the snakes when we were on foot, often surprising them along the river.

We did not find West Coast garter snakes in many localities but did see them regularly near the Río Cuchujaqui crossing south of Alamos. One August day in 1985, Susan Anderson, Peter Warren, and Cecil Schwalbe were snorkeling just below the crossing and saw seven garter snakes within an hour along about 1 km of stream. We caught and held four of the snakes overnight. Two of the snakes were gravid and gave birth that evening. We weighed, measured, and released all of the snakes the next morning, convinced that we had discovered the best method of collecting that snake. We have snorkeled for West Coast garter snakes at least a half-dozen times since then, with absolutely no success. Excessively turbid water from rainstorms limited visibility, or no snakes were found.

Rivers are the places to see more fishes, including the native cichlids *(Cichlosoma beani)* and several exotic species, such as large-mouth bass *(Micropterus salmoides)*, and tilapia *(Tilapia* spp.). Big-eyed leopard frogs *(Rana magnaocularis)* and Mexican mud turtles *(Kinosternon integrum)* are found both along rivers and in smaller streams. Couch's spadefoots *(Scaphiopus couchii)*, sabinal frogs *(Leptodactylus melanonotus)*, Great Plains narrowmouth toads *(Gastrophryne olivacea)*, and blackneck garter snakes can be found at the Río Cuchujaqui, but they are much more likely to live near smaller streams.

Ponds provide habitats for several species of amphibians and reptiles. The Alamos mud turtle *(Kinosternon alamosae)* prefers pond habitats near Alamos, whereas the Mexican mud turtle is more likely to live in streams in the region. These two mud turtles share some morphological and ecological traits with the two mud turtle species in Arizona. The Alamos mud turtle and the yellow mud turtle (K. flavescens) both have immaculate yellow faces and prefer pond habitats. The Mexican mud turtle and the Sonoran mud turtle (K. sonoriense) both have streaked faces and are more commonly found in streams, although they may also reside in ponds. Many of the anurans are primarily pond breeders, including the impressive Mexican leaf frog, Mexican tree frog (Smilisca baudini), Great Plains narrowmouth toad, Couch's spadefoot, Sonoran Desert toad, lowland burrowing treefrog (Pternohyla fodiens), sheep frog, and Sinaloa toad (Bufo mazatlanensis). Lowland burrowing treefrogs will also occasionally breed in streams. Red-spotted toads (B. punctatus) breed most often in streams but will also breed in ponds.

Cliffs and rocky outcrops are homes to rock specialists. These species include lyre snakes (Trimorphodon biscutatus and T. tau), Sonoran leaf-toed geckos (Phyllodactylus homolepidurus), Nelson's spiny lizards (Sceloporus nelsoni), and spinytail iguanas (Ctenosaura hemilopha), which also live in trees with cavity shelters.

FOLKLORE, HERPETOLOGICAL MYTHS, AND COMMERCE

Some of the people we have encountered who live in the numerous, small, isolated villages in the Sierra Madre Occidental are incredible naturalists. They have a vast store of knowledge about which plants to eat, which plants to use for medicinal purposes, which animals live in which habitats, what eats what, etc. Some individuals can distinguish four or five species of oaks or frogs and toads that live in their area. Finding a local naturalist or especially keen observer was always a boon to our collecting efforts. People were eager to help and proud to demonstrate their knowledge. However, even with such a rich understanding of their natural world, many people in Sonora, like residents in most countries, perpetuate myths surrounding amphibians and especially reptiles.

One of the most enduring and widespread myths in Mexico is that of the *ajolote*, the local name for several species of skinks (*Eumeces* spp.) in Sonora, two of which live in the Alamos area. Skinks are shiny, quick-moving lizards with very small legs that forage in surface litter during the day. Supposedly the smooth scales and tiny limbs allow an ajolote to crawl into a human rectum, where it lodges itself. We have never seen skinks seek refuge in any orifice of another animal, human or otherwise. Also called *ajolotes*, the mole lizards (*Bipes* spp.) from Baja California and Michoacán, which look like earthworms with two front legs, are also feared because of alleged similar burrowing behavior. *Ajolote* is also used in parts of Chihuahua to refer to salamanders (Thomas Van Devender, pers. comm. 1999).

Geckos, locally called *salamanquesas*, are believed to have an extremely venomous bite. These are totally harmless lizards that have gotten a bad rap. Several local people warned us not to handle them. A group became extremely alarmed when one of us acted as if he had been bitten by a salamanquesa and suddenly passed out. They seemed to delight in the joke when the truth came out and we could not coax the gecko to bite us. They did not, however, offer their fingers for the experiment.

The brown vine snake *(Oxybelis aeneus)* is extremely long and slender, a five-footer that at mid-body is only as wide as a human pinkie. Locally known as the *chicotera*, this snake supports an elaborate myth. Allegedly the chicotera will slither into a barn or corral, wrap its coils around the legs of a cow like piggin strings in a calf-roping contest, and suckle the cow's teats for milk. None of the three vine snakes to which we offered cow's milk in a bowl attempted to drink.

The pichecuate *(Agkistrodon bilineatus)* is greatly feared throughout its range, which is the TDF from southern Sonora to Guatemala, on the Yucatán Peninsula, and in Tamaulipas. It is called *cantíl* in the eastern part of its range. Although in reality the pichecuate is a pit viper with a fairly nasty bite, it is alleged to have almost supernatural powers, supposedly being able to leap 4 or 5 m through the air, sting with its tail (which does have a terminal spur), and cause almost instant death when it bites. People who survive a pichecuate bite, and most do, are often revered in their villages and proudly show off their scars to those who ask about their encounter.

Camaleones or horned lizards (*Phrynosoma* spp.) are reputed to squirt blood from their eyes to deter predators. This "myth" is actually true for several species, including the regal horned lizard *(P. solare)* found near Alamos. The reaction is strongest in response to attacks by canids. Naive dogs, kit foxes, or coyotes that were squirted in the mouth by Texas horned lizards *(P. cornutum)* would no longer attack them (Middendorf and Sherbrooke 1992; Wade Sherbrooke, pers. comm. 1998). Separating out myth from truth is one of the enjoyable pursuits in biology.

During the past several decades, herpetologists and herpetoculturists (those who keep and breed amphibians and reptiles in captivity, usually for sale) have hired local people to collect animals for them. In the late 1970s and early 1980s, the "Snake Lady of Alamos," as she became widely known in herpetological circles, sold from her *abarrote* (small shop) near the cemetery reptiles that her husband and neighborhood children had collected. She occasionally had as many as 20 turtles for sale at one time: a hodgepodge of desert tortoises, spotted box turtles, painted wood turtles, and mud turtles held in large barrels for potential buyers. Beaded lizards were favored fare because of their demand—mostly illegal—and the ease with which they could be kept in captivity on a diet of chicken eggs and water. Species that we saw for sale also included whipsnakes, speckled racers, parrot snakes, milk snakes, green rat snakes, indigo snakes, lyre snakes, rattlesnakes, western coral snakes, leaf-nosed snakes, and spinytail iguanas.

One of the species most often for sale was the filetail ground snake, a beautiful, small snake less than half a meter in length that is banded with variable numbers of brilliant red, yellow, and black "triads" and has a rasplike tail. Some of our friends were so enamored with this delicate, gorgeous creature that they paid the Snake Lady her $3-per-snake fee (1980 price index) and then released the snakes at the rock wall at the cemetery.

The Snake Lady still operates her abarrote but is out of the reptile business. One of her customers, a known black marketeer of reptiles in Arizona, finally got his comeuppance. After purchasing several beaded lizards in Alamos and driving to Nogales, Sonora, he was overcome with fatigue (and a bad drug habit) and parked his truck by the side of the road to sleep before crossing the border. His vehicle attracted the attention of the authorities, who seized his contraband, placed him under arrest, and interrogated him. Officials visited the Snake Lady and persuaded her to retire from reptile commerce.

Conservation

The greatest threats to the amphibians and reptiles of the Alamos area are the growth of the human population and the destruction and development of important habitats for human use. The devastating effects of the massive conversion of thornscrub and TDF communities into buffelgrass pastures for cattle are described elsewhere in this volume. Highland areas in the Sierra de Alamos and elsewhere are being used to grow marijuana, cocaine, and poppies for heroin. Although this may have local impacts on areas that are cleared to plant crops, growing drugs probably does not have such a negative effect on the herpetofauna because collectors, especially gringos, are discouraged from snooping around such enterprises.

For most species, the sale of reptiles on local black markets probably does not affect the population. However, the intensive collection of long-lived species, such as turtles and beaded lizards, can have negative, even catastrophic, effects on populations.

Alamos is a beautiful town. Its status as a national historical site requires that all new building construction be done in the old-style architecture of the late seventeenth and early eighteenth centuries, when the Sierra de Alamos was the hub of silver-mining activity for northwestern Mexico. Current economic prospects are based largely on tourism. We hope that the growth of the human resident and tourist population can be slowed so it does not destroy the unique habitats that support such a rich biota. Outlying areas are already threatened by development of trailer parks, camping areas, and birding stops.

ACKNOWLEDGMENTS

We are grateful to the many people who have provided important observations, assistance, and data over the years, including Ab Abercrombie, Guillermo Acosta, Susan Anderson, Randy Babb, Jody Lee Duek, George Ferguson, Mary Gilbert, Steve Hale, Peter Holm, Horatio Hurtado, Phil Jenkins, Alice Karl, Neal Krug, Howard Lawler, Brent Martin, Paul Martin, Pete Mayne, Stephanie Meyer, Paul Moler, Roy Murray, Bruce Palmer, Steve Prchal, Adam and Ethan Schwalbe, Dave Stejskal, Tom and Tim Van Devender, Peter Warren, Delia and Lou Weitzman, and Betsy Wirt. Tom Van Devender provided valuable comment on the chapter; he, Ana Lilia Reina Guerrero, and Alberto Búrquez also shared their knowledge of Spanish common names. We thank Pat Jenks for the generous access to his house in Alamos as a base of operations. For assistance on the checklist of mammals, we thank Randy Babb, Len Cockrum, Yar Petryszyn, Ronnie Sidner, Tom Van Devender, and Elvira Rojero, ecologist with the Instituto del Medio Ambiente y el Desarrollo Sustentable del Estado de Sonora. We sincerely apologize to anyone we may have inadvertently omitted.

Appendix 6.1. Amphibians of the Alamos, Sonora, Area

Scientific Name[a]	Common Name[a]	Habitat[b]	Abundance & Status[c]
BUFONIDAE			
Bufo alvarius	Sonoran Desert toad, Colorado River toad, sapo, sapo grande, sapo verde	TS, TDF	C
Bufo cognatus	Great Plains toad, sapo	TS	R
Bufo kelloggi	Little Mexican toad, sapito	TS	R

Scientific Name[a]	Common Name[a]	Habitat[b]	Abundance & Status[c]
Bufo marinus	Cane toad, sapo, sapo grande	TS	R
Bufo mazatlanensis	Sinaloa toad, sapo	TS, TDF, OW	A
Bufo punctatus	Red spotted toad, sapo	TS, TDF, OW	A
HYLIDAE			
Hyla arenicolor	Canyon treefrog, rana	Rocky s in TDF, TS	U
Pachymedusa dacnicolor	Mexican leaf frog, rana verde	TDF	C
Pternohyla fodiens	Lowland burrowing treefrog, rana	TS, TDF	U
Smilisca baudini	Mexican treefrog, rana, rana trepadora	TDF	U
LEPTODACTYLIDAE			
Eleutherodactylus augusti	Barking frog, rana, rana amarilla labradora	TS, TDF	U
Leptodactylus melanonotus	Sabinal frog, ranita	S, R in TS, TDF	LC
MICROHYLIDAE			
Gastrophryne olivacea	Great Plains narrowmouth toad, sapito	TS, TDF	U
Hypopachus variolosus	Sheep frog, sapito	TDF	R[d]
PELOBATIDAE			
Scaphiopus couchii	Couch's spadefoot, sapo	TS, TDF	A
RANIDAE			
Rana forreri	Sinaloan leopard frog, rana, rana del zacate	P	R[e]
Rana magnaocularis	Big-eyed leopard frog, rana	S, SP, R in TDF, TS	LC

Sources: Bogert and Oliver (1945), Flores-Villela (1993a), Heringhi (1969), Smith and Taylor (1966), and observations and collections by the authors and associates dating from the 1940s.

[a] Scientific names follow Collins (1997), Flores-Villela (1993a), or Stebbins (1985). Common names for the most part follow local usage in Sonora, Evans and Garcia (1998), Liner (1994), or Stebbins (1985). English common names are listed first, followed by Spanish common names.

[b] Habitats include only those used by species in the Alamos area. Habitats are listed in the order in which the species is most often found there. OW = oak woodland; P = ponds (often *represos*, or earthen stock tanks); R = riverine; S = stream; SP = spring; TDF = tropical deciduous forest; TS = foothills or coastal thornscrub.

[c] A = abundant (many individuals or groups found, typically over broad areas); C = common (several individuals or groups found, usually when conditions are appropriate); L = locally (used with letter C to indicate that several individuals or groups are present in suitable habitats); U = uncommon (species present but not always found); R = rare (species present but seldom found).

[d] The only report of sheep frogs in Sonora was a breeding population in a pond west of Alamos (Wiewandt et al. 1972). That pond has been altered as part of a chicken-ranching operation, and no sheep frogs have been seen in Sonora since that time.

[e] The Sinaloan leopard frog is locally common near the coast in northern Sinaloa and southern Sonora but does not normally live in TDF. A population was apparently introduced into represos in the Alamos area and observed by the authors in the 1980s. None have been seen there since 1990.

Appendix 6.2. Reptiles of the Alamos, Sonora, Area

Scientific Name[a]	Common Name[a]	Habitat[b]	Abundance & Status[c]
TESTUDINES			
BATAGURIDAE			
Rhinoclemmys pulcherrima	Painted wood turtle, tortuga de agua, tortuga pinta	TDF, S in summer	U
EMYDIDAE			
Terrapene nelsoni	Spotted box turtle, tortuga, juanon, tortuga de la caja	TDF, OW	U
Trachemys scripta	Slider, tortuga de agua	R in TDF, TS	U
KINOSTERNIDAE			
Kinosternon alamosae	Alamos mud turtle, tortuga de agua	P in TDF, TS	LC
Kinosternon integrum	Mexican mud turtle, tortuga de agua	S, R, P in TS, TDF	C
TESTUDINIDAE			
Gopherus agassizii	Desert tortoise, tortuga, tortuga del monte	TS, TDF	U
SQUAMATA			
SAURIA			
ANGUIDAE			
Elgaria kingii	Madrean alligator lizard, lagartija, cachora	OW, POF, TDF	U
EUBLEPHARIDAE			
Coleonyx fasciatus	Black banded gecko, salamanquesa	TDF	R
Coleonyx variegatus	Western banded gecko, salamanquesa	TS	R
GEKKONIDAE			
Phyllodactylus homolepidurus	Sonoran leaf-toed gecko, salamanquesa	C, M in TDF, TS	C
HELODERMATIDAE			
Heloderma horridum	Mexican beaded lizard, escorpión	TDF, TS	U
Heloderma suspectum	Gila monster, escorpión	TS	R
IGUANIDAE			
Ctenosaura hemilopha	Mainland spinytail iguana, iguana	TS, TDF	C
PHRYNOSOMATIDAE			
Callisaurus draconoides	Zebratail lizard, perrita	TS, TDF	C
Dipsosaurus dorsalis	Desert iguana, iguana	TS	R
Holbrookia maculata	Lesser earless lizard, lagartija, cachorita	TS, TDF	C

Appendix 6.2. (Continued)

Scientific Name[a]	Common Name[a]	Habitat[b]	Abundance & Status[c]
Phrynosoma solare	Regal horned lizard, camaleón	TS, TDF	U
Sceloporus clarkii	Clark's spiny lizard, cachora, cachorón	TDF, TS, OW	LC
Sceloporus horridus	Mexican spiny lizard, cachorón	TDF	U
Sceloporus magister	Desert spiny lizard, cachorón	TS	R
Sceloporus nelsoni	Nelson's spiny lizard, cachora	TDF	C
Urosaurus bicarinatus	Tropical tree lizard, lagartija	TDF	U
Urosaurus ornatus	Tree lizard, lagartija	TS, TDF	C
POLYCHRIDAE			
Norops nebulosus	Clouded anole, lagartija	TDF	C
SCINCIDAE			
Eumeces parviauriculatus	Little-eared skink, ajolote	TDF, POF	U
Eumeces tetragrammus	Four-lined skink, ajolote	TDF	U
TEIDAE			
Cnemidophorus costatus	Western Mexican whiptail, huico	TDF	C
SERPENTES			
BOIDAE			
Boa constrictor	Boa constrictor, corua	TDF, S in TS	U
COLUBRIDAE			
Chilomeniscus cinctus	Banded sand snake, culebra, víbora	TDF, TS	U
Dryadophis cliftoni	Lizard eater, lagartijera	TDF	R
Drymarchon corais	Western indigo snake, babatuco, culebra azul, culebra prieta	TDF, R in TS	U
Drymobius margaritiferus	Speckled racer, culebra, víbora	TDF	R
Elaphe triaspis	Green rat snake, culebra verde	TDF, TS	U
Gyalopion quadrangulare	Thornscrub hook-nosed snake, coralillo, viboríta, culebríta	TDF, TS	C
Hypsiglena torquata	Night snake, culebra, víbora	TDF, TS	U

Scientific Name[a]	Common Name[a]	Habitat[b]	Abundance & Status[c]
Imantodes gemmistratus	Blunthead tree snake, cordelilla de la escamuda, víbora, culebra	TDF	U
Lampropeltis getula	Common kingsnake, culebra	TS, TDF	R
Lampropeltis triangulum	Milk snake, coralillo	TDF	U
Leptodeira splendida	Splendid cat-eyed snake, víbora, culebra	TDF	R
Leptophis diplotropis	Pacific Coast parrot snake, culebra verde, güirotillera	TDF	U
Masticophis bilineatus	Sonoran whipsnake, chirrionero, chicotera, alicante, alicantre	TDF, OW, POF	U
Masticophis flagellum	Coachwhip, chirrionero, chicotera, alicante, alicantre	TS, TDF	U
Masticophis mentovarius	Neotropical whipsnake, chirrionero, chicotera, alicante, alicantre	TDF, R in TS, OW	U
Oxybelis aeneus	Brown vine snake, güirotillo, bejuquillo, chicotera	TDF, TS	U
Phyllorhynchus browni	Saddled leafnose snake, culebra, culebrita, víbora	TDF	C
Pituophis melanoleucus	Gopher snake, víbora sorda	TDF, TS	U
Pseudoficimia frontalis	False ficimia, culebra, víbora	TDF	R
Rhinocheilus lecontei	Longnose snake, coralillo	TDF, TS	C
Salvadora hexalepis	Western patchnose snake, víbora, culebra	TS, TDF	U
Sonora aemula	Filetail ground snake, coralillo, víbora, culebra	TDF	LC
Sympholis lippiens	Mexican short-tailed snake, culebra, víbora	TDF	R
Tantilla yaquia	Yaqui black-headed snake, culebra, víbora	TDF, TS	U
Thamnophis cyrtopsis	Blackneck garter snake, culebra del agua	S, R in TS, TDF, OW	C
Thamnophis eques	Mexican garter snake, culebra del agua	MA, S in OW, TS, TDF	U

Appendix 6.2. (Continued)

Scientific Name[a]	Common Name[a]	Habitat[b]	Abundance & Status[c]
Thamnophis valida	West Coast garter snake, culebra del agua	R in TDF	U
Trimorphodon biscutatus	Lyre snake, víbora, culebra	C in TDF, TS	LC
Trimorphodon tau	Mexican lyre snake, víbora, culebra	C in TDF, TS	U
CROTALIDAE			
Agkistrodon bilineatus	Mexican moccasin, pichecuate	TDF	R
Crotalus basiliscus	Mexican West Coast rattle-snake, cascabel, víbora de cascabel	TDF into TS	U
Crotalus tigris	Tiger rattlesnake, cascabel, víbora de cascabel	TS into TDF	R
ELAPIDAE			
Micruroides euryxanthus	Western coral snake, coralillo	TDF, TS, OW	R
Micrurus distans	West Mexican coral snake, coral, coralillo	TDF, TS	R
LEPTOTYPHLOPIDAE			
Leptotyphlops humilis	Western blind snake, culebrilla ciega, culebra, viboríta	TDF, TS, OW	R

Sources: Bogert and Oliver (1945), Flores-Villela (1993a), Heringhi (1969), Smith and Taylor (1966), and observations and collections by the authors and associates dating from the 1940s.

[a] Scientific names follow Collins (1997), Flores-Villela (1993a), or Stebbins (1985). Common names for the most part follow local usage in Sonora, Evans and Garcia (1998), Liner (1994), or Stebbins (1985). English common names are listed first, followed by Spanish common names.

[b] Habitats include only those used by species in the Alamos area. Habitats are listed in the order in which the species is most often found there. C = cliffs or rocky outcrops; M = mines or caves; MA = marshes *(ciénegas)*; OW = oak woodland; P = ponds (often represos); POF = pine-oak forest; R = riverine; S = stream; TDF = tropical deciduous forest; TS = foothills or coastal thornscrub.

[c] C = common (several individuals or groups found, usually when conditions are appropriate); L = locally (used with letter *C* to indicate that several individuals or groups are present in suitable habitats); U = uncommon (species present but not always found); R = rare (species present but seldom found).

Appendix 6.3. Amphibians and Reptiles That Live in Tropical Deciduous Forest in Sonora

Scientific Name[a]	Habitat[b]	Abundance & Status[c]
AMPHIBIANS		
CAUDATA		
AMBYSTOMATIDAE		
Ambystoma rosaceum[d] (E)	s in TDF, OW, POF, CF	C
ANURA		
BUFONIDAE		
Bufo alvarius	DS, TS, TDF	C
Bufo mazatlanensis (E)	TDF, OW	A
Bufo punctatus	TS, TDF, OW, G, DS, POF, CF	A
HYLIDAE		
Hyla arenicolor	Rocky s in TS, DS, G, OW, TDF, POF	U
Pachymedusa dacnicolor (E)	TDF, springside and streamside in TS	C
Pternohyla fodiens	TS, TDF, DS	U
Smilisca baudini	TDF	U
LEPTODACTYLIDAE		
Eleutherodactylus augusti	C in TS, TDF, OW	U
Leptodactylus melanonotus	S, R in TS, TDF	LC
MICROHYLIDAE		
Gastrophryne olivacea	TS, TDF, DS, G	U
Hypopachus variolosus	TDF	R[e]
PELOBATIDAE		
Scaphiopus couchii	TS, TDF, G, DS	A
RANIDAE		
Rana magnaocularis (E)	S, SP, R in TS, TDF, OW	LC
Rana pustulosa[f] (E)	s in TDF	U
Rana tarahumarae[g]	s in TDF, TS, OW, POF	U
REPTILES		
TESTUDINES		
BATAGURIDAE		
Rhinoclemmys pulcherrima	TDF, seasonally in s	U
EMYDIDAE		
Terrapene nelsoni (E)	TDF, OW	U
Trachemys scripta	R in TDF, TS	U
KINOSTERNIDAE		
Kinosternon alamosae (E)	P in TDF	LC
Kinosternon integrum	S, R, P in TS, TDF	C
TESTUDINIDAE		
Gopherus agassizii	TS, TDF, G	U

Appendix 6.3. (Continued)

Scientific Name[a]	Habitat[b]	Abundance & Status[c]
SQUAMATA		
SAURIA		
EUBLEPHARIDAE		
Coleonyx fasciatus (E)	TDF	R
Coleonyx variegatus	TS, DS, TDF	C (R in TDF)
GEKKONIDAE		
Phyllodactylus homolepidurus (E)	C, M in TDF, TS	C
HELODERMATIDAE		
Heloderma horridum	TDF, TS	U
Heloderma suspectum	TS	R
IGUANIDAE		
Ctenosaura hemilopha (E)	TS, TDF	C
PHRYNOSOMATIDAE		
Callisaurus draconoides	TS, TDF, DS	C
Holbrookia maculata	TS, TDF, G, OW	C
Phrynosoma ditmarsi[h] (E)	OW, TDF	R
Phrynosoma solare	DS, TS, TDF	U
Sceloporus clarkii	OW, TS, TDF, DS	LC
Sceloporus horridus (E)	TDF	U
Sceloporus nelsoni (E)	S, R in TDF	C
Urosaurus bicarinatus (E)	TDF	U
Urosaurus ornatus	TS, TDF, DS, G, OW, POF	C
POLYCHRIDAE		
Norops nebulosus (E)	TDF	C
SCINCIDAE		
Eumeces parviauriculatus (E)	TDF, POF, OW	U
Eumeces tetragrammus	TDF, S, R in OW, G	U
TEIDAE		
Cnemidophorus costatus (E)	TDF, TS	C
SERPENTES		
BOIDAE		
Boa constrictor	TDF, SP in TS	U
COLUBRIDAE		
Chilomeniscus cinctus	DS, TDF, TS	U (R in TDF)
Dryadophis cliftoni (E)	TDF	R
Drymarchon corais	TDF, R in TS	U
Drymobius margaritiferus	TDF	R
Elaphe triaspis	TS, TDF, OW	U
Gyalopion quadrangulare	TDF, TS	LC
Hypsiglena torquata	TDF, TS, OW, G	U
Imantodes gemmistratus	TDF	U

Scientific Name[a]	Habitat[b]	Abundance & Status[c]
Lampropeltis getula	S in DS, G, TS, TDF	C (R in TDF)
Lampropeltis triangulum	TDF	U
Leptodeira splendida (E)	TDF	R
Leptophis diplotropis (E)	TDF	U
Masticophis bilineatus	TDF, OW, POF	C (U in TDF)
Masticophis flagellum	DS, TS, TDF, G	C (U in TDF)
Masticophis mentovarius	TDF, R in TS	U
Oxybelis aeneus	TDF, TS	U
Phyllorhynchus browni	TDF, TS, G	U (C in TDF)
Pituophis melanoleucus	TDF, TS, G, DS	C (U in TDF)
Pseudoficimia frontalis	TDF	R
Rhinocheilus lecontei	DS, TDF, TS, G	C
Salvadora hexalepis	DS, TS, TDF	C (U in TDF)
Sonora aemula (E)	TDF	LC
Sympholis lippiens (E)	TDF	R
Tantilla yaquia	OW, TS, TDF	U
Thamnophis cyrtopsis	S, R in G, TS, OW, TDF, DS	LC
Thamnophis eques	MA, S in G, OW, TS, TDF	U
Thamnophis valida (E)	R in TDF	U
Trimorphodon biscutatus	C in TDF, TS	U (LC in TDF)
Trimorphodon tau (E)	C in TDF	U
CROTALIDAE		
Agkistrodon bilineatus	TDF	R
Crotalus basiliscus (E)	TDF, TS	U
Crotalus tigris	TS, DS, TDF	LC (R in TDF)
ELAPIDAE		
Micruroides euryxanthus	DS, TDF, TS, OW	R
Micrurus distans (E)	TDF, TS	R
LEPTOTYPHLOPIDAE		
Leptotyphlops humilis	TDF, TS, OW, G, DS	R

Sources: Bogert and Oliver (1945), Flores-Villela (1993a), Smith and Taylor (1966), and observations and collections by the authors and associates dating from the 1940s.

[a] Scientific names follow Collins (1997), Flores-Villela (1993a), or Stebbins (1985). English and Spanish common names are listed in footnotes here only for species that are not included in appendix 6.1 or 6.2. (E) indicates that the species is endemic to Mexico.

[b] Habitats include all those used by the species throughout Sonora, not just in the Alamos area. Habitats are listed approximately in the order in which the species is most often found there. C = cliffs or rocky outcrops; CF = coniferous forest; DS = desertscrub; G = grassland; M = mines or caves; MA = marshes (ciénegas); OW = oak woodland; P = ponds (often represos); POF = pine-oak forest; R = riverine; S = stream; SP = spring; TDF = tropical deciduous forest; TS = foothills or coastal thornscrub.

Appendix 6.3. (Continued)

---

^{d} Tarahumara salamander, salamandra Tarahumara.

^{e} The only report of sheep frogs in Sonora was a breeding population in a pond west of Alamos (Wiewandt et al. 1972). That pond has been altered as part of a chicken-ranching operation, and no sheep frogs have been seen in Sonora since that time.

^{f} Cascade frog, rana de cascada.

^{g} Tarahumara frog, rana de Tarahumara.

^{h} Rock horned lizard, camaleón.

APPENDIX 6.4

Mammals of the Alamos, Sonora, Area

The list of mammals of southern Sonora is impressively diverse, with at least 81 species confirmed for the Alamos area; each of these species is marked with an asterisk (*) in the table. Tropical mammal species contribute significantly to this richness, especially among the bats. This list has been constructed from the handful of published reports on the area (Burt 1938; Cockrum and Bradshaw 1963; Goldman 1951; Hall 1981), William Caire's dissertation (Caire 1978), and a preliminary list of mammals of the Sierra de Alamos–Río Cuchujaqui area (Rojero 1998).

Many people have shared their observations of interesting mammals in the Alamos area, including several herpetologists, who have a propensity to inspect roadkills. Occasionally photographs have been useful in confirming the presence of species not yet vouchered by a physical specimen. It would be interesting to see how many more mammal species could be verified in the Alamos area if mammals were surveyed as intensively as amphibians, birds, and reptiles have been during the past few decades.

Seven species of mammals are believed to live in the Alamos area, but no local photographs or voucher specimens have been obtained; each of these species is marked with a dagger (†) in the table. Thirteen additional species may live in the area, although the habitat near Alamos is marginal at best for many of them; each of these species is marked with a double dagger (‡) in the table. The jaguarundi, for example, has been reported as likely to live in the Alamos area in several of our principal references (Caire 1978; Goldman 1951; Hall 1981; Rojero 1998). Although a few unconfirmed sightings have been reported in Sonora and Arizona, no specimens or photographs of jaguarundis exist for either state.

Appendix 6.4. Mammals of the Alamos, Sonora, Area

Scientific Name[a]	Common Name[a]	Source[b]
MARSUPIALIA		
DIDELPHIDAE		
Didelphis virginiana	Virginia opossum, tlacuache, tacuachi	B, C, H
Marmosa canescens	Grayish mouse-opossum, tacuachito	C, H
INSECTIVORA		
SORICIDAE		
Notiosorex crawfordi	Desert shrew, musaraña	C, G, H, R
CHIROPTERA		
DESMODONTIDAE		
Desmodus rotundus	Vampire bat, vampiro, murciélago vampiro	C, H, R
EMBALLONURIDAE		
Balantiopteryx plicata	Peters' bat, murciélago	B, C, H, R
MOLOSSIDAE		
Eumops perotis	Western mastiff bat, murciélago	C, H
Eumops underwoodi	Underwood's mastiff bat, murciélago	C, H, R
‡ *Molossus molossus*	Pallas' mastiff bat, murciélago	H
Nyctinomops aurispinosus	Free-tailed bat, murciélago	C, H, R
Nyctinomops femorosaccus	Pocketed free-tailed bat, murciélago	C, H, R
Nyctinomops macrotis	Big free-tailed bat, murciélago	C, H, R
Tadarida brasiliensis	Brazilian free-tailed bat, murciélago	B, C, CB, H, R
MORMOOPIDAE		
Mormoops megalophylla	Peters' ghost-faced bat, murciélago	B, C, CB, H, R
Pteronotus davyi	Davy's naked-backed bat, murciélago	C, CB, H, R
Pteronotus parnellii	Parnell's mustached bat, murciélago	B, C, CB, H, R
Pteronotus personatus	Wagner's mustached bat, murciélago	C, H, R
NATALIDAE		
Natalus stramineus	Mexican funnel-eared bat, murciélago	C, CB, H, R

Appendix 6.4. (Continued)

Scientific Name[a]	Common Name[a]	Source[b]
PHYLLOSTOMATIDAE		
Artibeus hirsutus	Hairy fruit-eating bat, murciélago	C, CB, H, R
‡ *Centurio senex*	Wrinkle-faced bat, murciélago	H
‡ *Chiroderma salvini*	Salvin's white-lined bat, murciélago	C, H
Choeronycteris mexicana	Mexican long-tongued bat, murciélago	B, C, H, R
‡ *Glossophaga commissarisi*	Commissaris' long-tongued bat, murciélago	H
Glossophaga soricina	Davis' long-tongued bat, murciélago	B, C, CB, H, R
Leptonycteris curasoae	Southern long-nosed bat, murciélago nariz grande	B, C, CB, R
Macrotus californicus	California leaf-nosed bat, murciélago	B, C, H
Macrotus waterhousii	Waterhouse's leaf-nosed bat, murciélago	C, H, R
Sturnira lilium	Yellow-shouldered bat, murciélago	C, CB, H, R
VESPERTILIONIDAE		
Antrozous pallidus	Pallid bat, murciélago, murciélago pálido	C, H, R
Eptesicus fuscus	Big brown bat, murciélago	C, H, R
‡ *Idionycteris phyllotis*	Allen's big-eared bat, murciélago	C
Lasiurus borealis	Eastern red bat, murciélago	B, C, H, R
Lasiurus cinereus	Hoary bat, murciélago	C, H, R
Lasiurus ega	Southern yellow bat, murciélago	C, H, R
Myotis auriculus	Southwestern myotis, murciélago	C, H, R
† *Myotis californicus*	California myotis, murciélago	C, H, R
Myotis fortidens	Cinnamon myotis, murciélago	C, H, R
† *Myotis occultus*	Occult bat, murciélago	R
† *Myotis thysanodes*	Fringed bat, murciélago	B, R
Myotis velifer	Cave myotis, murciélago	B, C, H, R
Myotis yumanensis	Yuma myotis, murciélago	B, C, H, R
Pipistrellus hesperus	Western pipestrelle, murciélago	C, H, R

Scientific Name[a]	Common Name[a]	Source[b]
‡*Plecotus mexicanus*	Mexican big-eared bat, murciélago	C, H
Plecotus townsendii	Townsend's big-eared bat, murciélago	C, H, R
Rhogeëssa parvula	Little yellow bat, murciélago	C, H, R
Rhogeëssa tumida	Central American yellow bat, murciélago	B
EDENTATA		
DASYPODIDAE		
Dasypus novemcinctus	Nine-banded armadillo, armadillo	C, H, R
LAGOMORPHA		
LEPORIDAE		
Lepus alleni	Antelope jackrabbit, liebre	B, C, G, H, R
Sylvilagus audubonii	Desert cottontail, conejo del desierto, conejito	B, C, G, H, R
Sylvilagus floridanus	Eastern cottontail, conejo del bosque, conejito	C, H, R
RODENTIA		
CRICETIDAE		
Baiomys taylori	Northern pygmy mouse, ratón	B, C, H, R
Neotoma albigula	White-throated packrat, rata de campo	B, C, G, H, R
Neotoma mexicana	Mexican packrat, rata de campo	B, G, H, R
Neotoma phenax	Sonoran packrat, tori, rata de campo	C, H, R
Onychomys torridus	Southern grasshopper mouse, ratón	B, C, CB, G, H, R
Peromyscus boylii	Brush mouse, ratón	B, C, H, R
Peromyscus eremicus	Cactus mouse, ratón	B, C, CB, H, R
Peromyscus merriami	Merriam's mouse, ratón	C, H, R
Reithrodontomys burti	Sonoran harvest mouse, ratón	C, H
Reithrodontomys fulvescens	Fulvous harvest mouse, ratón	B, C, G, H
Sigmodon arizonae	Arizona cotton rat, rata	C, H, R
ERETHIZONTIDAE		
Erethizon dorsatum	Common porcupine, puerco espín	C, H
GEOMYIDAE		
Thomomys bottae	Botta's pocket gopher, tuza, ratón	B, C, G, H, R

Scientific Name[a]	Common Name[a]	Source[b]
HETEROMYIDAE		
Chaetodipus artus	Narrow-skulled pocket mouse, ratón	B, C, G, H, R
Chaetodipus baileyi	Bailey's pocket mouse, ratón	C, H
Chaetodipus goldmani	Goldman's pocket mouse, ratón	B, C, CB, G, H, R
‡ *Chaetodipus penicillatus*	Desert pocket mouse, ratón	C, H, R
Chaetodipus pernix	Sinaloan pocket mouse, ratón	B, C, G, H, R
Dipodomys merriami	Merriam's kangaroo rat, rata canguro	B, C, G, H, R
Liomys pictus	Painted spiny pocket mouse, ratón	B, C, G, H, R
MURIDAE		
† *Mus musculus*	House mouse, ratón	B, C
† *Rattus norvegicus*	Norway rat, rata	C
Rattus rattus	Black rat, rata	B, C
SCIURIDAE		
Sciurus colliaei	Collie squirrel, ardilla	B, C, H, R
Sciurus nayaritensis	Apache squirrel, Mexican fox squirrel, ardilla	C, H, R
‡ *Spermophilus madrensis*	Sierra Madre mantled ground squirrel, ardilla	C, H
† *Spermophilus tereticaudus*	Round-tailed ground squirrel, ardilla	B, H
Spermophilus variegatus	Rock squirrel, ardilla de las rocas	B, C, H, R
‡ *Tamias dorsalis*	Cliff chipmunk, chichimoco, ardillita	C, H
CARNIVORA		
CANIDAE		
Canis latrans	Coyote, coyote	B, C, H, R
‡ *Canis lupus*	Gray wolf, lobo	B, C, H
Urocyon cinereoargenteus	Common gray fox, zorra gris	C, H, R
FELIDAE		
‡ *Herpailurus yagouarundi*	Jaguarundi, leoncillo	C, G, H, R
Leopardus pardalis	Ocelot, ocelote	B, C, G, H, R
Leopardus wiedii	Margay, tigrillo, margay	C, H, R; photo of skin
Lynx rufus	Bobcat, gato montes	B, C, G, H, R

Scientific Name[a]	Common Name[a]	Source[b]
Panthera onca	Jaguar, tigre, jaguar	B, C, H, R
Puma concolor	Mountain lion, león, puma, león de la sierra	C, H, R
MUSTELIDAE		
Conepatus mesoleucus	Common hog-nosed skunk, zorrillo	B, C, H, R
Lutra annectens	Southern river otter, nutria del río	C
Mephitis macroura	Hooded skunk, zorrillo	B, C, H, R
‡ *Mephitis mephitis*	Striped skunk, zorrillo listado, rayado	C, H
‡ *Mustela frenata*	Long-tailed weasel, comadreja	C, H, R
Spilogale gracilis	Western spotted skunk, zorrillo pinto	C, R
Taxidea taxus	American badger, tejón	B, C, G, H, R
PROCYONIDAE		
Bassariscus astutus	Ringtail, cacomixtle	C, H, R
Nasua nasua	Coati, chulo, cholugo, coati	B, C, H, R
Procyon lotor	Common raccoon, mapache, batepi, lavador	B, C, H, R
URSIDAE		
† *Ursus americanus*	Black bear, oso negro	C, H, R
ARTIODACTYLA		
CERVIDAE		
Odocoileus virginianus	White-tailed deer, venado cola blanca	B, C, H, R
SUIDAE		
Pecari tajacu	Collared peccary, jabalí, cochi javalín	C, H, R

[a] Scientific and English common names follow Hall (1981), Jones et al. (1997), or Nowak (1991) in most cases. Spanish common names follow Evans and Garcia (1998) or Rojero (1998), with additional names provided by the Arizona-Sonora Desert Museum and field investigators.

[b] Sources are as follows: B = Burt (1938); C = Caire (1978); CB = Cockrum and Bradshaw (1963); G = Goldman (1951); H = Hall (1981); R = Rojero (1998).

*Mammals that have been confirmed in the scientific literature by specimen voucher or in identifiable photographs.

† Mammals that probably live in the area. Although the sources indicate that the species range is near Alamos, no specimens or photographs of the species have been taken in that area.

‡ Mammals that possibly live in the area. Although the sources indicate that the species range is near Alamos, no specimens or photographs of the species have been taken in that area.

REFERENCES

Bogert, C. M., and J. A. Oliver. 1945. A preliminary analysis of the herpetofauna of Sonora. *Bulletin of the American Museum of Natural History* 83:297–426.

Bowden, C., J. W. Dykinga, and P. S. Martin. 1993. *The secret forest.* Albuquerque: University of New Mexico Press.

Burt, W. H. 1938. *Faunal relationships and geographic distribution in Sonora, Mexico.* Miscellaneous Publications No. 39. Ann Arbor: Museum of Zoology, University of Michigan.

Caire, W. 1978. The distribution and zoogeography of the mammals of Sonora, Mexico, vols. 1 and 2. Ph.D. dissertation, University of New Mexico, Albuquerque.

Campbell, J. A., and W. W. Lamar. 1989. *The venomous reptiles of Latin America.* Ithaca, N.Y.: Comstock Publishing Associates.

Cockrum, E. L., and G.V.R. Bradshaw. 1963. Notes on mammals from Sonora, Mexico. *American Museum Novitates* 2138:2–10.

Collins, J. T. 1997. *Standard common and current scientific names for North American amphibians and reptiles,* 4th ed. Herpetological Circular No. 25. Athens, Ohio: Society for the Study of Amphibians and Reptiles.

Evans, D., and J. Garcia. 1998. *Desert life: A vocabulary.* Tucson: Arizona-Sonora Desert Museum.

Fa, J. E., and L. M. Morales. 1993. Patterns of mammalian diversity in Mexico. In *Biological diversity of Mexico: Origins and distribution,* eds. T. P. Ramamoorthy, R. A. Bye Jr., A. Lot, and J. Fa, 319–361. Oxford: Oxford University Press.

Flores-Villela, O. 1993a. *Herpetofauna Mexicana: Annotated list of the species of amphibians and reptiles of Mexico, recent taxonomic changes, and new species.* Special Publication No. 17. Pittsburgh: Carnegie Museum of Natural History.

———. 1993b. Herpetofauna of Mexico: Distribution and endemism. In *Biological diversity of Mexico: Origins and distribution,* eds. T. P. Ramamoorthy, R. A. Bye Jr., A. Lot, and J. Fa, 253–280. Oxford: Oxford University Press.

Goldman, E. A. 1951. *Biological investigations in Mexico.* Miscellaneous Collections No. 115. Washington, D.C.: Smithsonian Institution.

Hall, E. R. 1981. *The mammals of North America.* New York: Wiley.

Heringhi, H. L. 1969. An ecological survey of the herpetofauna of Alamos, Sonora, Mexico. Master's thesis, Arizona State University, Tempe.

Hilton, J. W. 1947. *Sonora sketch book.* New York: Macmillan.

Jones, C., R. S. Hoffmann, D. W. Rice, M. D. Engstrom, R. D. Bradley, D. J. Schmidly, C. A. Jones, and R. J. Baker. 1997. *Revised checklist of North American mammals north of Mexico, 1997.* Occasional Papers No. 173. Lubbock: Museum of Texas Tech University.

Liner, E. M. 1994. *Scientific and common names for the amphibians and reptiles of Mexico in English and Spanish.* Herpetological Circular No. 23. Athens, Ohio: Society for the Study of Amphibians and Reptiles.

Middendorf, G. A., III, and W. C. Sherbrooke. 1992. Canid elicitation of blood-squirting in a horned lizard *(Phrynosoma cornutum). Copeia* 1992:519–527.

Nowak, R. M. 1991. *Walker's mammals of the world,* 5th ed., 2 vols. Baltimore: Johns Hopkins University Press.

Rojero, E. 1998. Listado preliminar de mamíferos del área Sierra de Álamos — Arroyo Cuchujaqui. Hermosillo, Sonora: Instituto del Medio Ambiente y el Desarrollo Sustentable del Estado de Sonora.

Schoenhals, L. C. 1988. *A Spanish-English glossary of Mexican flora and fauna*. Mexico City: Summer Institute of Linguistics.

Smith, H. M., and E. H. Taylor. 1945. *An annotated checklist and key to the snakes of Mexico*. Bulletin No. 187. Washington, D.C.: U.S. National Museum.

———. 1948. *An annotated checklist and key to the amphibia of Mexico*. Bulletin No. 194. Washington, D.C.: U.S. National Museum.

———. 1950. *An annotated checklist and key to the reptiles of Mexico exclusive of the snakes*. Bulletin No. 199. Washington, D.C.: U.S. National Museum.

———. 1966. *Herpetology of Mexico: Annotated checklists and keys to the amphibians and reptiles. A reprint of Bulletins 187, 194, and 199 of the U.S. National Museum with a list of subsequent taxonomic innovations*. Ashton, Md.: Eric Lundberg.

Stebbins, R. C. 1985. *A field guide to western reptiles and amphibians*, 2d ed. Boston: Houghton Mifflin.

Turner, R. M., and D. E. Brown. 1982. Tropical-subtropical desertlands: 154.1 Sonoran desertscrub. In *Biotic communities of the American Southwest — United States and Mexico*, ed. D. E. Brown, special issue of *Desert Plants* 4:181–221.

Van Devender, T. R., J. I. Mead, and A. M. Rea. 1991. Late Quaternary plants and vertebrates from Picacho Peak, Arizona. *Southwestern Naturalist* 36:320–324.

Wiewandt, T. A., C. H. Lowe, and M. W. Larson. 1972. Occurrence of *Hypopachus variolosus* (Cope) in the short-tree forest of southern Sonora, Mexico. *Herpetologica* 28:162–164.

7

BIRDS OF THE TROPICAL
DECIDUOUS FOREST OF THE
ALAMOS, SONORA, AREA

Stephen M. Russell

The traveler heading to Alamos from Navojoa encounters several changes in the landscape. Initially one sees many clearings, poultry farms, and often swarms of vultures. Any persisting original vegetation appears sparse and low, similar to scrubby desert growth over much of Sonora and even the deserts of the U.S. Southwest. But as the first hills are traversed several kilometers east of Navojoa, the plant growth becomes increasingly more dense. Trees seem larger and more abundant. Their height increases, and they ultimately overshadow the tall spindly cacti. If the trip is made in the humid months from June to September, the hillsides are a mass of green, in marked contrast to the arid features that prevail the rest of the year. One has entered the northern end of the tropical deciduous forest (TDF), a narrow belt or corridor of forest that extends southward along the edge of the Pacific Ocean to Costa Rica.

The vegetational changes in the 50-km transect from Navojoa to Alamos are highly visible, the consequence of the presence of many tropical plant species that reach their northernmost geographic limits in this region. Less conspicuous but equally dramatic are the changes in animal life. With a little good fortune, one may see a dozen large blue, white, and black birds fly across the highway, trailing greatly elongated tail feathers. These are Black-throated Magpie-Jays (*Calocitta colliei*), relatives of the all-black crows and ravens (*Corvus* spp.) that inhabit so many parts of the world. The Black-throated Magpie-Jay is the most obvious indicator of a bird fauna that is just as distinctive as the plant community in which it lives.

In southeastern Sonora, 351 species of birds have been documented (appendix 7.1), a remarkably high number for any locality that lacks a coastline or extensive wetlands. No other area in Sonora has such a high diversity. About 150 of these species (appendix 7.2) are characteristic of the TDF region, although a substantial number are also present in other plant communities. If one spent a day or two observing birds near Alamos, it would not be difficult to see 100 bird species. Of these, 12 would be ones that have never been found in the United States; another 25 would be birds that barely reach southern Arizona.

Sources of Information

The status of birds in Sonora was summarized in a book by A. J. van Rossem in 1945 and was based heavily upon specimen collections made during the previous 75 years. Alamos was the center of considerable early collecting; many nests and eggs were taken in the Güirocoba area in the 1930s and 1940s (see fig. 3.1). My own fieldwork in the area began in 1965, and I have made numerous short trips there. In 1969 Alden published a book on finding birds in western Mexico based on his many trips as a tour leader. His writing and natural history tours resulted in Alamos becoming a popular destination for bird watchers. Short (1974) published a paper on the summer birds of the Alamos region. Over the past 25 years, many people have made trips to southeastern Sonora and have kept useful accounts of their observations. Gale Monson and I recently wrote *The Birds of Sonora* (Russell and Monson 1998), based upon our own field observations, specimens in many museums, and the field notes of competent observers.

The observations and conclusions expressed in this paper are based upon information in Russell and Monson (1998), supplemented by my fieldwork in the Alamos region. The birds of the TDF in southern Sonora await intensive study. No studies provide basic information on exact habitat preferences. There are no quantitative data that express densities of breeding, migrating, or wintering birds. My assignments of bird species to various habitat types are largely subjective.

Appendix 7.1 lists the number of localities in which each bird species is known within the defined region. Sixty-eight localities at elevations from 100 m to 1,630 m are represented; most of these are at low elevations. At some localities, records of only a few birds exist, but other places have had extensive fieldwork by many individuals during all seasons. For example, the Broad-billed Hummingbird *(Cynanthus latirostris)* has been reported from 42 of the 68 localities—more localities than for any other species. It should not be assumed, however, that this hummingbird does not live in other areas.

The extent of data from each locality is biased in many ways. Field observers tend to pay more attention to rarities than to common birds, and they are more likely to see conspicuous species than reclusive ones. The Broad-billed Hummingbird is a colorful species and barely reaches the United States, thus it is of great interest. It is more likely to be reported than a Dusky-capped Flycatcher *(Myiarchus tuberculifer)*, which has been recorded at 40 locations. Given the absence of any systematic census of birds in southern Sonora, the locality data combined with the elevational range provide a crude but useful index of the distribution of each species.

It is difficult to be precise in defining the limits of the TDF community. Appendix 7.2 identifies species according to a general classification of their habitats: forest, secondary growth areas, open areas (including pastures, cultivated land, fencerows, and abandoned fields), and riparian growth areas. Where undisturbed, riparian growth consists of tall trees that typically overhang streams. Most waterbirds are excluded from this list; the presence of forest has little to do with their presence or absence in the region. Thus, herons, egrets, ducks, and kingfishers do not appear on the list, although they are often present on bodies of water in forested areas. The Bare-throated Tiger-Heron *(Tigrisoma mexicanum)* is included, however, because it is limited to stream banks with adjacent forest. Many species of birds are so-called habitat generalists and may occupy any of several different habitats for foraging or nesting. Other species are narrowly restricted to a single community. Species that nest in forest alone and are intolerant of even secondary growth areas are at risk when their habitat is altered or fragmented. Birds that live in both forest and secondary growth communities are much more tolerant of human-induced changes.

Based on residency status, the birds that constitute the TDF avifauna reflect several different categories. Permanent residents remain in one small area throughout the year. Summer residents are species that undergo a seasonal migration, returning from some southern nonbreeding wintering area to nest in the Alamos area, usually during the period from April to October. Most migrants are species that move long distances, flying south in the autumn and north in the spring. A small number of Alamos area birds are altitudinal migrants; i.e., they breed in the high mountains of eastern Sonora and western Chihuahua and migrate down to the foothills for the winter. Many migrants spend the winter months in the TDF, where they compete with permanent residents for resources for 7 or 8 months.

Significance of the TDF Avifauna

The TDF of western Central America is a relatively narrow corridor extending from southern Sonora to northwestern Costa Rica. For the most part, it is restricted to Pacific slope drainage. In it have evolved a number of distinctive birds as well as other animals and plants. The highest concentration of species in the TDF is in the area from Oaxaca to El Salvador (Stiles 1983). Sonora is at the northern end of this area and has the fewest species, yet they contribute significantly to the region's bird diversity.

The most complete species list for a TDF habitat in western Mexico is from Chamela, Jalisco (Arizmendi et al. 1990; Hutto 1992). Hutto (1992) included a list of 98 species in disturbed and undisturbed TDF habitats. Of these, 18 do not reach Sonora, but 64 are forest birds in the Alamos area (included in appendix 7.2).

Of the 100 most widely distributed species (on the basis of number of localities; appendix 7.1), 21 have distributional ranges that extend northward from the Alamos area barely into Arizona. The deciduous forest corridor, although it gradually disappears north of Alamos, has provided the route northward.

Many birds in deciduous forest areas are migrants (table 7.1). Of the 120 species of birds that live in forests (listed in appendix 7.2 in the habitats designated as F, R, FR, FRS, FS, and FRSO), 35% (42 species) are migrants. Some or many individuals of 36 of the 42 migrant species are also present through the winter. The Costa's Hummingbird *(Calypte costae)* has been found only in winter in undisturbed forest; all other species may be present in forest, tall secondary growth, and riparian forest areas. Migrant birds in southern Sonora have not been studied to determine if any have unique requirements. Some North American migratory birds are known to have specific habitat needs while migrating and during the winter.

TDF Avifauna

DECIDUOUS FOREST SPECIES

Eighteen species are characteristic of forest habitat (category F in appendix 7.2) but not other communities. All presumably nest there (or did so in the past), except the Costa's Hummingbird, which is present only in winter. Only two of these species live in the United States.

Individual Military Macaws *(Ara militaris)* may move over large areas each

Table 7.1. Number of Bird Species Characteristic of Habitat Types

Habitat[a]	T[b]	P B+BP[c]	S B+BP[d]	? B[e]	M[f]	W[g]	Loc[h]
F	18	11	4	2	0	1	9.9
R	5	2	0	1	0	2	8.8
FR	15	8	6	1	0	0	17.4
FRS	57	20	5	0	6	26	20.4
FS	13	9	2	0	0	2	17.5
FRSO	12	9	0	0	0	3	30.1
S	1	0	0	0	0	1	16.0
SR	6	1	2	0	0	3	17.3
SO	14	6	2	1	0	5	12.6
SRO	9	4	2	0	1	2	16.9
O	1	0	1	0	0	0	5.0
A	1	1	0	0	0	0	2.0
Total	152	71	24	5	7	45	

Source: Russell and Monson (1998), supplemented by Russell's field experience in Alamos area.
Note: This table summarizes the species listed in appendix 7.2 by habitats.
[a] F = deciduous forest; R = riparian forest; S = secondary growth areas; O = open areas; A = aerial (forages in air over all habitats).
[b] Total species characteristic of habitat type(s).
[c] Permanent residents breeding (b = breeding confirmed; bp = breeding presumed).
[d] Summer residents breeding (b = breeding confirmed; bp = breeding presumed).
[e] Not known if likely to be breeding in region.
[f] Migrants from more northern breeding areas that spend winter farther south.
[g] Migrants that spend winter in region.
[h] Average number of localities in which species in group have been noted.

day but typically avoid regions where human disturbance is high. The macaw's distribution in Sonora includes areas closely associated with deciduous forest that are only 150 km from the U.S. border. The Lilac-crowned Parrot (*Amazona finschi*) lives in both deciduous and coniferous forests in Sonora. Flights to feeding areas may take it over cleared land. Its numbers have declined in the last few years, especially in deciduous forest areas. It reaches its northern limits in southeastern Sonora.

The Lesser Roadrunner *(Geococcyx velox)* differs only slightly in appearance from its widespread and well-known relative, the Greater Roadrunner *(G. californianus)*. The Greater Roadrunner is rarely found in the forest but is common in open and secondary growth areas. The Lesser Roadrunner lives only in deciduous forest but is most often found where the slopes are steep, typically rocky, and well drained and where the vegetation is sparser.

Two species of small owls reach their northern limits in deciduous forest. Both the Vermiculated Screech-Owl *(Otus guatemalae)* and the Central American Pygmy-Owl *(Glaucidium griseiceps)* are rare in Sonora, and little is known about them. The Plain-capped Starthroat *(Heliomaster constantii)* is an uncommon hummingbird. Unlike most species in the deciduous forest group, it has been recorded a few times in southern Arizona, although it is known to nest only in extreme southeastern Sonora and southward.

The only species in this group that is present as a migrant or wintering bird is the Costa's Hummingbird, which breeds in the Sonoran Desert. Small numbers of female plumaged birds, a category that includes adult females and young birds of both sexes, are present in winter in deciduous forest in the Alamos area. Adult males winter farther north.

The Russet-crowned Motmot *(Momotus mexicanus)*, a member of an exclusively Neotropical family, ranges northward through the deciduous forest to the Alamos area. It is present only in summer, when it nests in holes in steep banks above streams.

Four deciduous forest species nest in tree cavities and forage on bark. Both Lineated and Pale-billed Woodpeckers *(Dryocopus lineatus* and *Campephilus guatemalensis)* are large-bodied species, the largest woodpeckers in western Mexico. (The Imperial Woodpecker *[Campephilus imperialis]*, which is larger, is probably extinct.) Although the Lineated Woodpecker is rare, it is regularly seen. The Pale-billed Woodpecker may have been extirpated from Sonora; it has not been seen there since 1950. The Neotropical family Dendrocolaptidae is represented at the northern edge of its range in Sonora by the Ivory-billed and White-striped Woodcreepers *(Xiphlorhynchus flavigaster* and *Lepidocolaptes leucogaster)*. The Ivory-billed Woodcreeper is rare in Sonora, and its numbers are apparently declining. The White-striped Woodcreeper is uncommon but widespread in deciduous forest. It extends upward into pine-oak forest and is limited to deciduous forest only at lower elevations.

Thirty-eight species of flycatchers (Tyrannidae) have been found in Sonora; two atypical members of the family are known only from deciduous forest. The rare Bright-rumped Attila *(Attila spadiceus)* inhabits deep canyons, usually along small streams. The Masked Tityra *(Tityra semifasciata)*, which is fairly common in many places south to Panama, is known in Sonora only from the extreme southeastern corner, close to the border of Chihuahua and Sinaloa. Van Rossem (1945) found several in May 1937, but they have not been found again in Sonora.

The Purplish-backed Jay *(Cyanocorax beecheii)* is a distinctive but uncommon resident of deciduous forest. It is a social species, typically congregating in small flocks that include young from previous seasons. The Spotted Wren *(Campylorhynchus gularis)*, a very close relative of the Cactus Wren *(C. brunneicapillus)*, is locally dispersed in deciduous forest upward into pine-oak woodland. Yellow-winged Caciques *(Cacicus melanicterus)* nested near Alamos in 1931 (van Rossem 1945) but have been seen only once subsequently (in 1982, south of Alamos on the Río Cuchujaqui).

The Short-tailed Hawk *(Buteo brachyurus)* is included in the list of deciduous forest species, but its distribution in Sonora is poorly known. It is a wide-ranging species that was first seen in the state in 1982. No nests have been located, although its numbers are increasing. Most observations have been of birds in deciduous forest, but three have been seen in pine-oak forest. Although it may ultimately become widespread in a variety of habitats in Sonora, its movement northward appears to have been through the deciduous forest corridor.

RIPARIAN FOREST SPECIES

The five species in this group (habitat category R in appendix 7.2) have geographic ranges that extend from South America northward only as far as Sonora. They are almost invariably found along streams in areas of deciduous forest. Three of these species breed in the high mountains of eastern Sonora and southwestern Chihuahua and migrate seasonally to lowland riparian areas.

The Bare-throated Tiger-Heron *(Tigrisoma mexicanum)* forages in shallow water and on the banks of streams. Nests are placed in tall riparian trees. They are often seen along the Río Cuchujaqui, sitting in a tall sabino or on a cliff face. The Great Black-Hawk *(Buteogallus urubitinga)* is rare in Sonora, but the five recorded sightings were all along rivers where there were tall trees. The species is so rare that it is not known if its numbers are declining.

The Tufted Flycatcher *(Mitrephanes phaeocercus)*, Orange-billed Nightingale-Thrush *(Catharus aurantiirostris)*, and White-throated Robin *(Turdus assimilis)* all nest in the mountains and migrate to wooded lowland valleys for the winter. The Tufted Flycatcher is a characteristic nesting bird in pine forests and is regularly seen along streams in the Alamos area in winter. The Orange-billed Nightingale-Thrush was first spotted in Sonora in 1965 (on the Río Cuchujaqui southeast of Alamos); this species is rare and rather poorly known. There is a small breeding population northeast of Alamos in the mountains and also in nearby Chihuahua. Presumably these are the birds that winter along small lowland streams. The White-throated Robin nests above 900 m in wooded canyon bottoms at the upper limits of deciduous forest and in adjacent pine-oak woodland. It moves into the lowlands in fall in search of fruit, such as figs. It is not conspicuous, but falling fruit may be an indication that robins are feeding in a fig tree.

DECIDUOUS AND RIPARIAN FOREST SPECIES

This category includes species characteristic of the interior of deciduous forest and of tall trees along streams, where they are most regularly found. Of the 15 species in this group (habitat category FR in appendix 7.2), 10 have been found in the United States (5 nesting regularly).

Large raptors are represented by the Common Black-Hawk *(Buteogallus anthracinus)* and Solitary Eagle *(Harpyhaliaetus solitarius)*. The Common Black-Hawk frequents riparian areas from Argentina to southern Arizona and is uncommon and local in the northern part of this range. The Solitary Eagle is rare throughout its range, especially in Mexico. Three nests have been found in Sonora, the latest in 1958 (Harrison and Kiff 1977; Russell and Monson 1998). Other than a possible observation in 1987, the species has not been seen in Sonora since 1958. It ranges upward into pine forest.

A large and spectacular cuckoo, the Squirrel Cuckoo *(Piaya cayana)* has a long graduated tail and bright rufous underparts. Its range extends from Argentina to Sonora, but it is an uncommon and reclusive resident of forest thickets. The Mottled Owl *(Ciccaba virgata)* is rare in Sonora. All records are from deciduous and riparian forests. One bird found west of San Bernardo was flushed from its roost on an hecho. The Violet-crowned Hummingbird *(Amazilia violiceps)* is a common nesting species in Sonora, and a few breed in southern Arizona. It is most frequently seen near streams. Northern populations are migratory.

Two cavity-nesting flycatchers, the Brown-crested Flycatcher *(Myiarchus tyrannulus)* and the Sulphur-bellied Flycatcher *(Myiodynastes luteiventris)*, are often conspicuous in areas with tall trees. Both have ranges that extend southward from Sonora, primarily through deciduous forest, to northwestern Costa Rica. To the north, the Brown-crested Flycatcher is widespread in the southwestern United States, even in pine-oak woodland. The Sulphur-bellied Flycatcher enters southern Arizona and New Mexico, where it occupies riparian areas in pine-oak woodland. Both flycatchers are migratory; the Sulphur-bellied spends its winters east of the Andes in Bolivia and Peru.

The Rose-throated Becard *(Pachyramphus aglaiae)*, which is also migratory (a few, however, winter in deciduous forest in the Alamos area), is common in summer. Its large globular nests often hang on high branches over water from southern Arizona to Costa Rica.

One of many species of robins, the Rufous-backed Robin *(Turdus rufopalliatus)* is a resident breeding bird in all deciduous and riparian forests. When not nesting, individuals flock and move through the Alamos area in search of small fruit, such as figs, which they regularly consume. Wandering individuals occasionally appear in southern Arizona. The Blue Mockingbird *(Melanotis caerulescens)* is poorly known, partly because of its secretive nature. It can be easily found only when it sings during the breeding season (May–July). Breeding pairs are uncommon but are found at many localities in deciduous forest, especially above 900 m in riparian areas. During the winter, birds from higher elevations move into lower riparian thickets in areas of deciduous forest and are extremely difficult to see.

The past and present status of the Golden Vireo *(Vireo hypochryseus)* in Sonora is uncertain. Seven specimens were taken in 1888 and another one in 1937 (van Rossem 1945). There are no recent records that are suitably documented. The species is considered to be resident in its range (south to Oaxaca), but it is possible that the Sonoran population is not only very rare but also migratory and present for only a short time in summer. The uncommon Yellow-green Vireo *(V. flavoviridis)* spends only 3 months in Sonora. It is a summer nesting species, returning from its wintering grounds in late May, nesting in July, and departing in August.

The western Mexico population of the Tropical Parula *(Parula pitiayumi)*, a tiny warbler, reaches its northern limit in the deciduous forest of Sonora. The species ranges from southern Texas to Argentina. Unlike other populations, which are sedentary, Sonoran birds are migratory. They return from wintering areas in late April, breed in June as the summer rains begin, and usually depart by the first week of September. The Yellow Grosbeak *(Pheucticus chrysopeplus)* has almost the same migratory pattern as the Tropical Parula. It arrives in mid-May, nests in June, and departs in early September. Both its bright coloration and its loud and pleasant song make it easy to find in May and June. Both the warbler and the grosbeak are most common in large trees along streams.

Although the Black-vented Oriole *(Icterus wagleri)* is found almost anywhere in deciduous forest, streams with adjacent palms provide the optimal habitat. The bird uses the palms for nesting, and large groups may roost in them during winter.

DECIDUOUS FOREST, RIPARIAN FOREST, AND SECONDARY GROWTH AREA SPECIES

Of the 152 species of birds that are associated with deciduous forest (appendix 7.2), 70 are most regularly found in deciduous forest, riparian forest, and secondary growth areas (habitat categories FRS and FS in appendix 7.2). As a group, these species are not habitat limited but range widely through the Alamos area. Seven species are summer residents, 28 are permanent residents, 6 are migrants that pass through Sonora twice a year, and 29 winter in Sonora but breed to the north. Only the 13 species that have never reached the border with Arizona or New Mexico are mentioned here.

A rare raptor in Sonora, the Crane Hawk *(Geranospiza caerulescens)* has been found nesting only in the Güirocoba area. Several nests of the Laughing Falcon *(Herpetotheres cachinnans)* were found in the 1940s, all near Güirocoba, but no birds have been seen in Sonora since 1949. This is a conspicuous bird and likely to be noted if present. It has probably been extirpated from the state. The Rufous-bellied Chachalaca *(Ortalis wagleri)* is uncommon but widespread in southeastern Sonora. Its call may be heard over a considerable distance, but individuals are infrequently seen. Its numbers appear to be stable, although it is often hunted. The common quail throughout the Alamos region is the Elegant Quail *(Callipepla douglasii)*.

Although large, the Red-billed Pigeon *(Columba flavirostris)* may be difficult to see when it sits silently in the tops of tall trees. Some individuals leave the forested nesting areas in autumn to forage on grain in agricultural regions to the west. The range of the common White-tipped Dove *(Leptotila verreauxi)* extends from Argentina to south Texas to deciduous forest in western Mexico. In Sonora it reaches 30°18'N latitude, at the extreme northern limit of the TDF.

The Mexican Parrotlet *(Forpus cyanopygius)* was once common in the Alamos area (Alden 1969) but is rarely seen now, except in the town of Alamos. Presumably its numbers have declined. It may be migratory because this tiny bird has never been seen from October through March outside of Alamos. Little is known about the biology of the species in the northern part of its range. The apparent sharp decline in its numbers indicates that the species is in need of study, if not help.

The White-fronted Parrot *(Amazona albifrons)* is a fairly common resident in the lowlands of southern Sonora, and a few individuals are occasionally found where pines are present. Its numbers appear to be stable. Inappropriately named, the Mangrove Cuckoo *(Coccyzus minor)* does not live in mangroves in Sonora but is an uncommon summer resident in patchy deciduous forest. All records of sightings are from June, July, and August, indicating that its nesting activity is closely tied to the rainy season. Certainly one of the most conspicuous birds in the Alamos region, the Black-throated Magpie-Jay sometimes forages quietly in dense trees and can be easily overlooked. It is a permanent resident.

Two closely related species of small wrens are residents of TDF thickets. The Sinaloa Wren *(Thryothorus sinaloa)* and the Happy Wren *(T. felix)* are similar in appearance and habitat preferences, but the Sinaloa Wren may be found in more xeric places and is more widespread. A pair was found in 1986 only 100 km south of the Arizona border in a wooded area that included some plants of TDF origin. The Scrub Euphonia *(Euphonia affinis)* is known only from two specimens taken at Alamos in March 1888 (van Rossem 1945) and individuals spotted in December 1982 and 1985 in the Sierra de Alamos. The range of the species extends southward in TDF to northwestern Costa Rica.

DECIDUOUS FOREST, RIPARIAN FOREST, SECONDARY GROWTH AREA, AND OPEN AREA SPECIES

Only 12 species fall into this habitat category (FRSO in appendix 7.2). All except the Sinaloa Crow *(Corvus sinaloae)* breed in the United States. This crow has a very limited distribution in western Mexico, from Sonora south to Colima. It is a social species; large flocks are often seen, usually in agricultural regions. The species has increased in both number and range as more land has been converted to irrigated agriculture. The remaining species in this category are common and widely distributed.

NONFOREST SPECIES IN THE TDF REGION

The 32 species in this group (habitat categories s, sr, sro, so, o, and a in appendix 7.2) are not typically found within undisturbed forest, but they are characteristic of secondary growth areas. Only the Bat Falcon *(Falco rufigularis)* and Common Pauraque *(Nyctidromus albicollis)* have not reached the southwestern United States by moving northward through western Mexico.

The Bat Falcon is rare in western Mexico, and it may no longer live in Sonora. Ten specimens and two clutches of eggs were collected in the Güirocoba area and one specimen was taken near Ciudad Obregón, all in the period from 1937 to 1949. I know of no other observations.

The Common Pauraque is extending its range northward. The first Sonoran birds were discovered in 1980 north of Ciudad Obregón. A nest with eggs was found in 1986. The only record in the Alamos area is of three calling birds on the Río Cuchujaqui southeast of Alamos in 1988.

Breeding Seasons

In a region where a long dry season is characteristic, birds would be expected to nest close to the beginning of the seasonal rains. Parent birds require a substantial protein source to feed their nestlings and ensure their growth; this is most readily available in insects. Young birds that have just left the nest and are learning to forage require readily obtainable food, as well as a rich protein source. The flush of insects associated with the onset of the summer rains in June fills these needs.

Table 7.2 summarizes the breeding season of TDF birds. Data are available on 92 of the 100 species believed to nest in TDF in the Alamos area. Table 7.2 indicates that most species begin nesting in May or June, immediately before the rains commence. The timing is such that insects are abundant when there are nestlings to be fed. Even more important, there are many insects when young birds are becoming independent and are learning to find their own food.

Population Changes in TDF Avifauna

After centuries of land-use practices that did not destroy the TDF, major changes are taking place. The clearing of TDF and the planting of buffelgrass (*Pennisetum ciliare*) will have severe impacts upon forest area birds. Not only is the extent of remaining forest being greatly reduced, but the resulting fragmentation is also detrimental to many species. Removal of a habitat may result in a complete loss of those species most characteristic of it. Although patches of forest may be left (*fragmentation*), many species require extensive tracts of habitat for their survival (Askins et al. 1990; Kattan et al. 1994; Robbins et al. 1989; Willson et al. 1994).

Table 7.2. Number of Bird Species Known (or Believed) to Breed in Tropical Deciduous Forest in the Alamos Region Each Month

SPECIES GROUPS	F	M	A	M	J	J	A	S	TOTAL[a]
Tiger-heron, falconiforms	1	2	8	10	9	7	1		12
Chachalaca, quail, pigeons, doves		2	3	5	8	8	3	2	8
Parrots			2	3	4	2			4
Cuckoos				2	6	6	4	1	6
Owls	1	1	2	7	6	3			7
Nightjars, swifts			1	2	3	4			4
Hummingbirds	1	2	3	3	3	1	1		3
Trogons, woodpeckers, woodcreepers			1	8	9	9	2		9
Flycatchers			3	7	10	8	2		11
Jays, wrens, gnatcatchers, robins			1	7	11	10	2		11
Mimids, vireos, warblers, finches			2	3	9	11	6	1	12
Icterines, House Finch			2	3	5	5	2		5
Total[b]	3	7	28	60	83	74	23	4	92

Sources: Russell and Monson (1998), Short (1974), and van Rossem (1945).
Note: Nesting is based upon observations of nests, the presence of young birds incapable of sustained flight, behavior (territoriality and courtship), and the condition of the gonads of birds prepared for specimens. Data from many years were combined, including some years that were very wet or very dry. The monthly totals are probably underestimated because most field observers have devoted little attention to the breeding behavior of birds. The species are listed in appendix 7.2.
[a] Column headings are as follows: F = February; M = March; A = April; M = May; J = June; J = July; A = August; S = September; Total = total number of species represented by data.
[b] Total number of species nesting each month.

Fragmentation also increases the impact of predators and brood parasites on nesting birds (Paton 1994) by increasing the extent of forest edge. Many predators and brood parasites are open area species that do not penetrate deeply into forests, and are thus not a threat when forest tracts are large. Two species of brood parasites, the Bronzed Cowbird *(Molothrus aeneus)* and the Brown-headed Cowbird *(Molothrus ater)*, are common and widespread in the Alamos area. They deposit their eggs in the nests of other species of birds, invariably curtailing the nesting success of the host. As more land is converted to grazing, cowbirds will proliferate to the detriment of TDF birds.

DECLINING SPECIES AND SPECIES OF
QUESTIONABLE STATUS

In the absence of long-term monitoring information that would provide density data, it is difficult to be precise in identifying species that are declining in southern Sonora. For species that have dropped in number or that have not been reported in many years, it is difficult to identify the cause. Several of the 17 species described as restricted to deciduous forest may no longer be present in Sonora or are declining at an alarming rate. The Lineated Woodpecker is declining in numbers. The Pale-billed Woodpecker has not been seen since 1950. Reports of the Ivory-billed Woodcreeper appear to be less frequent, but possibly they have always been rare. The Masked Tityra has been found in Sonora only once (in 1937) but may have been nesting at that time. The Yellow-winged Cacique nested near Alamos in 1931 but has since disappeared from the state. The Bright-rumped Attila is rare, and its past and present status is uncertain.

The Solitary Eagle has always been rare, with no positive records since 1958. Only five sightings of the Great Black-Hawk are well documented from Sonora. No Laughing Falcon has been reported since 1949, but there were several nests in the 1940s. The Bat Falcon disappeared from the state in 1949.

There is only one record of the Green Parakeet *(Aratinga holochlora)*. The Mexican Parrotlet, once common, is declining. The Lilac-crowned Parrot is also declining in number. The Military Macaw, although apparently holding its own, is vulnerable to habitat disturbance. Many macaw species in Latin America have endangered or threatened status, brought about by habitat disruption and human hunting pressure (for the captive bird trade). The Golden Vireo is extremely rare in Sonora.

It is likely that the clearing of TDF has spurred the decline of many of these species. The selective cutting of large trees may be detrimental for large woodpeckers and cavity-nesting species. Large raptors and game birds are especially vulnerable to human impact (Robinson and Wilcove 1989; Thiollay 1989). As TDF is being rapidly cleared from Sonora to Costa Rica, those who have the power to stop the destruction of this unique ecosystem should focus their attention on the potential demise of innumerable plants and animals.

INCREASING SPECIES

A number of bird species are expanding their range northward and increasing in number. The Short-tailed Hawk has used the deciduous forest corridor in its northward expansion. Although it is mostly restricted to forest, the species may ultimately become dispersed in other habitats.

Most species that are increasing are ones that inhabit open country, particularly seed-eating birds, such as doves and finches. The Alamos area will probably always have many birds. The number of individual birds will likely rise, but there will be an associated decline in diversity. High bird species diversity is an indicator of a healthy environment, one that is also capable of sustaining human life.

Appendix 7.1. Birds of the Alamos, Sonora, Area

Scientific Name[a]	Common Name[a]	N[b]	Min. El.[c]	Max. El.[d]	Med. El.[e]	Abun./ Stat.[f]
PODICIPEDIFORMES						
PODICIPEDIDAE						
Tachybaptus dominicus	Least Grebe	3	260	480	330	R/Pb
Podilymbus podiceps	Pied-billed Grebe	1	260	—	—	R/W
Podiceps nigricollis	Eared Grebe	1	100	—	—	R/W
PELECANIFORMES						
PELECANIDAE						
Pelecanus erythrorhynchos	White Pelican	1	290	—	—	R/W
PHALACROCORACIDAE						
Phalacrocorax brasilianus	Neotropic Cormorant	6	155	480	305	U/P
Phalacrocorax auritus	Double-crested Cormorant	2	155	260	—	U/P
ANHINGIDAE						
Anhinga anhinga	Anhinga	1	260	—	—	R/P
CICONIIFORMES						
ARDEIDAE						
Tigrisoma mexicanum	Bare-throated Tiger-Heron	7	220	600	270	U/Pb
Ardea herodias	Great Blue Heron	16	100	480	220	C/Pbp
Ardea alba	Great Egret	8	100	480	260	U/P
Egretta thula	Snowy Egret	4	155	260	230	U/P

Scientific Name[a]	Common Name[a]	N[b]	Min. El.[c]	Max. El.[d]	Med. El.[e]	Abun./ Stat.[f]
Bubulcus ibis	Cattle Egret	2	250	520	—	U/P
Butorides virescens	Green Heron	10	160	480	290	U/Pbp
Nycticorax nycticorax	Black-crowned Night-Heron	2	155	260	—	U/Pbp
CICONIIDAE						
Mycteria americana	Wood Stork	2	260	300	—	R/P
CATHARTIDAE						
Coragyps atratus	Black Vulture	37	125	1,630	320	C/Pb
Cathartes aura	Turkey Vulture	39	125	1,630	320	C/Pbp
ANSERIFORMES						
ANATIDAE						
Dendrocygna autumnalis	Black-bellied Whistling-Duck	7	125	205	160	C/Pbp
Anas strepera	Gadwall	2	155	260	—	U/W
Anas americana	American Wigeon	1	260	—	—	U/W
Anas platyrhynchos	Mallard	1	350	—	—	R/W
Anas discors	Blue-winged Teal	1	390	—	—	U/W
Anas cyanoptera	Cinnamon Teal	2	260	1,580	—	U/W
Anas clypeata	Northern Shoveler	1	155	—	—	U/W
Anas acuta	Northern Pintail	1	350	—	—	U/W
Anas crecca	Green-winged Teal	3	155	390	260	U/W
Aythya valisineria	Canvasback	1	155	—	—	R/W
Aythya collaris	Ring-necked Duck	1	250	—	—	U/W
Aythya affinis	Lesser Scaup	2	155	260	—	U/W
Bucephala albeola	Bufflehead	1	155	—	—	R/W
Bucephala clangula	Common Goldeneye	1	400	—	—	R/W
Lophodytes cucullatus	Hooded Merganser	1	160	—	—	R/W
Mergus merganser	Common Merganser	2	160	210	—	U/W
Oxyura jamaicensis	Ruddy Duck	1	155	—	—	U/W
FALCONIFORMES						
ACCIPITRIDAE						
Pandion haliaetus	Osprey	4	125	210	170	U/W
Elanus leucurus	White-tailed Kite	2	260	350	—	R/P

Scientific Name[a]	Common Name[a]	N[b]	Min. El.[c]	Max. El.[d]	Med. El.[e]	Abun./ Stat.[f]
Haliaeetus leucocephalus	Bald Eagle	1	155	—	—	A/W
Circus cyaneus	Northern Harrier	2	260	390	—	U/W
Accipiter striatus	Sharp-shinned Hawk	14	150	1,500	335	U/Sbp C/W
Accipiter cooperii	Cooper's Hawk	14	155	1,500	385	U/Sbp C/W
Accipiter gentilis	Northern Goshawk	2	1,380	1,580	—	R/Pbp R/W
Geranospiza caerulescens	Crane Hawk	6	205	1,000	395	R/Pb
Asturina nitidus	Gray Hawk	30	125	1,420	335	C/Pb
Buteogallus anthracinus	Common Black-Hawk	17	185	1,500	310	C/Pb
Buteogallus urubitinga	Great Black-Hawk	3	250	390	260	A/?
Parabuteo unicinctus	Harris's Hawk	2	155	400	—	R/P
Harpyhaliaetus solitarius	Solitary Eagle	5	400	1,320	910	R/?b
Buteo brachyurus	Short-tailed Hawk	7	125	1,630	390	U/P?
Buteo swainsoni	Swainson's Hawk	1	350	—	—	U/M
Buteo albicaudatus	White-tailed Hawk	1	390	—	—	R/P?
Buteo albonotatus	Zone-tailed Hawk	13	155	1,580	620	U/Sb
Buteo jamaicensis	Red-tailed Hawk	29	125	1,630	400	C/Pb C/W
FALCONIDAE						
Caracara plancus	Crested Caracara	25	125	550	250	C/Pb
Herpetotheres cachinnans	Laughing Falcon	2	400	1,580	—	R/Pb
Falco sparverius	American Kestrel	17	100	1,480	320	C/Pbp C/W
Falco columbarius	Merlin	1	390	—	—	R/W
Falco rufigularis	Bat Falcon	1	400	—	—	R/Pb
Falco peregrinus	Peregrine Falcon	6	125	390	200	U/Pbp
GALLIFORMES						
CRACIDAE						
Ortalis wagleri	Rufous-bellied Chachalaca	20	125	1,000	390	U/Pb
PHASIANIDAE						
Meleagris gallopavo	Wild Turkey	2	880	1,000	—	U/Pbp
Callipepla douglasii	Elegant Quail	26	125	1,600	270	C/Pbp
Cyrtonyx montezumae	Montezuma Quail	6	800	1,630	1,250	U/Pb

Scientific Name[a]	Common Name[a]	N[b]	Min. El.[c]	Max. El.[d]	Med. El.[e]	Abun./ Stat.[f]
GRUIFORMES						
RALLIDAE						
Porzana carolina	Sora	1	350		—	U/Pbp
Fulica americana	American Coot	3	100	240	155	C/Pbp
CHARADRIIFORMES						
CHARADRIIDAE						
Charadrius vociferus	Killdeer	9	100	480	220	A/W
RECURVIROSTRIDAE						
Himantopus mexicanus	Black-necked Stilt	1	400	—	—	U/W
SCOLOPACIDAE						
Tringa melanoleuca	Greater Yellowlegs	4	155	260	245	U/W
Tringa solitaria	Solitary Sandpiper	2	220	260	—	C/W
Actitis macularia	Spotted Sandpiper	14	155	1,500	245	U/W
Calidris minutilla	Least Sandpiper	2	155	240	—	A/W
Limnodromus scolopaceus	Long-billed Dowitcher	1	155	—	—	U/W
Gallinago gallinago	Common Snipe	3	155	520	400	A/W
LARIDAE						
Larus delawarensis	Ring-billed Gull	1	155	—	—	A/W
Sterna caspia	Caspian Tern	2	100	155	—	C/Pbp
COLUMBIFORMES						
COLUMBIDAE						
Columba livia	Rock Dove	1	590	—	—	C/Pb
Columba flavirostris	Red-billed Pigeon	17	125	600	275	U/Pbp
Columba fasciata	Band-tailed Pigeon	6	1,000	1,630	1,540	C/Pb
Zenaida asiatica	White-winged Dove	40	125	1,630	350	C/Pb C/W
Zenaida macroura	Mourning Dove	25	125	1,600	380	C/Pb
Columbina inca	Inca Dove	30	155	1,630	350	C/Pb
Columbina passerina	Common Ground-Dove	35	100	1,380	320	C/Pb
Columbina talpacoti	Ruddy Ground-Dove	2	260	390	—	R/Pbp
Leptotila verreauxi	White-tipped Dove	40	125	1,630	400	C/Pb

Appendix 7.1. (Continued)

Scientific Name[a]	Common Name[a]	N[b]	Min. El.[c]	Max. El.[d]	Med. El.[e]	Abun./ Stat.[f]
PSITTACIFORMES						
PSITTACIDAE						
Aratinga holochlora	Green Parakeet	1	620	—	—	A/?
Ara militaris	Military Macaw	17	125	1,630	400	U/Sb
Rhynchopsitta pachyrhyncha	Thick-billed Parrot	2	460	580	—	R/P?
Forpus cyanopygius	Mexican Parrotlet	13	195	1,000	400	U/Sbp
Amazona albifrons	White-fronted Parrot	27	125	1,420	350	C/Pb
Amazona finschi	Lilac-crowned Parrot	20	250	1,630	610	R/Pb
CUCULIFORMES						
CUCULIDAE						
Coccyzus americanus	Yellow-billed Cuckoo	15	125	1,600	350	C/Sb
Coccyzus minor	Mangrove Cuckoo	5	195	600	350	U/Sb
Piaya cayana	Squirrel Cuckoo	19	195	1,000	400	U/Pb
Geococcyx velox	Lesser Road-runner	9	280	1,000	475	R/Pb
Geococcyx californianus	Greater Roadrunner	20	100	1,630	320	C/Pb
Crotophaga sulcirostris	Groove-billed Ani	10	135	400	250	U/Sb
STRIGIFORMES						
TYTONIDAE						
Tyto alba	Barn-Owl	4	270	390	350	U/Pbp
STRIGIDAE						
Otus kennicottii	Western Screech-Owl	19	150	910	350	C/Pb
Otus trichopsis	Whiskered Screech-Owl	5	400	1,630	1,580	C/Pb
Otus guatemalae	Vermiculated Screech-Owl	3	380	1,000	400	R/Pb
Bubo virginianus	Great Horned Owl	16	125	1,500	255	C/Pbp
Glaucidium gnoma	Northern Pygmy-Owl	8	270	1,630	1,250	U/Pbp
Glaucidium griseiceps	Central American Pygmy-Owl	6	400	1,000	750	R/Pb

Scientific Name[a]	Common Name[a]	N[b]	Min. El.[c]	Max. El.[d]	Med. El.[e]	Abun./ Stat.[f]
Glaucidium brasilianum	Ferruginous Pygmy-Owl	22	125	1,250	365	U/Pb
Micrathene whitneyi	Elf Owl	11	125	520	390	C/Sbp
Ciccaba virgata	Mottled Owl	11	155	1,000	475	U/Pb
Strix occidentalis	Spotted Owl	1	1,580	—	—	A/P
CAPRIMULGIFORMES						
CAPRIMULGIDAE						
Chordeiles acutipennis	Lesser Night-hawk	9	125	520	260	C/Sbp
Chordeiles minor	Common Nighthawk	5	155	1,600	1,500	C/Sbp
Nyctidromus albicollis	Common Pauraque	1	260	—	—	A/?
Phalaenoptilus nuttallii	Common Poorwill	1	390	—	—	A/?
Nyctiphrynus mcleodii	Eared Poorwill	3	880	1,600	1,320	R/P
Caprimulgus ridgwayi	Buff-collared Nightjar	20	125	1,580	395	C/Sb
Caprimulgus vociferus	Whip-poor-will	6	260	1,630	1,540	U/Sbp
APODIFORMES						
APODIDAE						
Cypseloides niger	Black Swift	2	1,500	1,600	—	R/?
Streptoprocne rutila	Chestnut-collared Swift	2	460	1,600	—	R/?
Streptoprocne semicollaris	White-naped Swift	2	100	800	—	R/?
Chaetura vauxi	Vaux's Swift	3	260	520	390	R/M
Aeronautes saxatalis	White-throated Swift	8	270	1,630	1,110	U/Pbp
TROCHILIDAE						
Cynanthus latirostris	Broad billed Hummingbird	42	125	1,630	385	C/Pb
Hylocharis leucotis	White-eared Hummingbird	12	390	1,630	1,380	U/Pbp
Amazilia beryllina	Berylline Hummingbird	16	195	1,630	775	C/Pbp
Amazilia rutila	Cinnamon Hummingbird	1	910	—	—	A/?
Amazilia violiceps	Violet-crowned Hummingbird	33	125	1,600	400	C/Pb

Scientific Name[a]	Common Name[a]	N^b	Min. El.[c]	Max. El.[d]	Med. El.[e]	Abun./ Stat.[f]
Lampornis clemenciae	Blue-throated Hummingbird	6	880	1,630	1,540	U/Pbp
Eugenes fulgens	Magnificent Hummingbird	5	880	1,600	1,500	U/Pbp
Heliomaster constantii	Plain-capped Starthroat	17	125	800	390	U/Pb
Archilochus alexandri	Black-chinned Hummingbird	5	220	480	275	U/M
Calypte anna	Anna's Hummingbird	1	390	—	—	R/?
Calypte costae	Costa's Hummingbird	4	260	440	335	U/W
Selasphorus platycercus	Broad-tailed Hummingbird	3	260	520	390	U/M
Selasphorus rufus	Rufous Hummingbird	5	250	1,420	400	U/M
Selasphorus sasin	Allen's Hummingbird	1	390	—	—	R/M
TROGONIFORMES						
TROGONIDAE						
Trogon elegans	Elegant Trogon	41	125	1,630	475	C/Pb
CORACIIFORMES						
MOMOTIDAE						
Momotus mexicanus	Russet-crowned Motmot	12	125	1,000	560	U/Sb
ALCEDINIDAE						
Ceryle alcyon	Belted King-fisher	11	100	480	240	U/W
Chloroceryle americana	Green Kingfisher	15	100	580	250	C/Pbp
PICIFORMES						
PICIDAE						
Melanerpes lewis	Lewis's Woodpecker	1	1,580	—	—	A/W
Melanerpes formicivorus	Acorn Wood-pecker	11	460	1,630	1,420	C/Pbp
Melanerpes uropygialis	Gila Woodpecker	36	125	800	315	C/Pb
Sphyrapicus thyroideus	Williamson's Sapsucker	2	260	1,500	—	R/W
Sphyrapicus varius	Yellow-bellied Sapsucker	1	260	—	—	R/W
Sphyrapicus nuchalis	Red-naped Sapsucker	6	213	1,500	325	U/W

Scientific Name[a]	Common Name[a]	N[b]	Min. El.[c]	Max. El.[d]	Med. El.[e]	Abun./ Stat.[f]
Picoides scalaris	Ladder-backed Woodpecker	33	125	800	350	C/Pb
Picoides villosus	Hairy Wood-pecker	1	1,500	—		R/P
Picoides stricklandi	Strickland's Woodpecker	11	400	1,630	1,480	U/Pb
Piculus auricularis	Gray-crowned Woodpecker	4	1,380	1,600	1,500	R/Pbp
Colaptes auratus	Northern Flicker	7	250	1,630	1,200	U/Pb U/W
Colaptes chrysoides	Gilded Flicker	14	125	400	240	C/Pb
Dryocopus lineatus	Lineated Woodpecker	11	195	1,000	500	R/Pbp
Campephilus guatemalensis	Pale-billed Woodpecker	6	390	1,000	610	R/Pb

PASSERIFORMES

DENDROCOLAPTIDAE

Scientific Name	Common Name	N	Min. El.	Max. El.	Med. El.	Abun./Stat.
Xiphlorhynchus flavigaster	Ivory-billed Woodcreeper	9	290	1,000	600	R/Pb
Lepidocolaptes leucogaster	White-striped Woodcreeper	17	125	1,630	1,380	U/Sbp

TYRANNIDAE

Scientific Name	Common Name	N	Min. El.	Max. El.	Med. El.	Abun./Stat.
Camptostoma imberbe	Northern Beardless-Tyrannulet	28	125	1,630	300	C/Pb
Mitrephanes phaeocercus	Tufted Flycatcher	16	240	1,600	470	U/Sb U/W
Contopus cooperi	Olive-sided Flycatcher	1	400	—	—	R/M
Contopus pertinax	Greater Pewee	21	240	1,630	670	C/Sb U/W
Contopus sordidulus	Western Wood-Pewee	11	260	1,630	620	C/Sb U/M
Empidonax traillii	Willow Flycatcher	5	220	480	260	U/M
Empidonax minimus	Least Flycatcher	1	390	—	—	A/W
Empidonax hammondii	Hammond's Flycatcher	9	240	1,380	520	U/M
Empidonax wrightii	Gray Flycatcher	12	150	520	300	C/W
Empidonax oberholseri	Dusky Flycatcher	16	150	1,380	385	C/W
Empidonax affinis	Pine Flycatcher	1	1,500	—	—	A/?
Empidonax difficilis	Pacific-slope Flycatcher	6	260	1,380	410	C/W
Empidonax occidentalis	Cordilleran Flycatcher	2	600	1,420	—	R/?

Scientific Name[a]	Common Name[a]	N[b]	Min. El.[c]	Max. El.[d]	Med. El.[e]	Abun./ Stat.[f]
Empidonax fulvifrons	Buff-breasted Flycatcher	8	240	1,480	335	U/Sbp U/W
Sayornis nigricans	Black Phoebe	27	155	1,600	350	C/Pb
Sayornis saya	Say's Phoebe	6	155	400	335	U/W
Pyrocephalus rubinus	Vermilion Flycatcher	17	100	1,500	270	C/Pb
Attila spadiceus	Bright-rumped Attila	5	260	1,320	440	R/Pbp
Myiarchus tuberculifer	Dusky-capped Flycatcher	40	125	1,630	400	R/Pb C/W
Myiarchus cinerascens	Ash-throated Flycatcher	14	150	1,380	440	R/Pb C/W
Myiarchus nuttingi	Nutting's Flycatcher	35	125	910	380	C/Pb
Myiarchus tyrannulus	Brown-crested Flycatcher	24	125	1,420	365	U/Sb
Pitangus sulphuratus	Great Kiskadee	3	240	480	260	U/Pbp
Myiozetetes similis	Social Flycatcher	8	100	480	305	U/Pbp
Myiodynastes luteiventris	Sulphur-bellied Flycatcher	24	125	1,630	440	C/Sb
Tyrannus melancholicus	Tropical Kingbird	16	125	520	230	C/Sb
Tyrannus vociferans	Cassin's Kingbird	15	125	1,600	305	C/Sb U/W
Tyrannus crassirostris	Thick-billed Kingbird	28	125	1,630	350	C/Sb
Tyrannus verticalis	Western Kingbird	8	220	1,600	390	U/M
Pachyramphus major	Gray-collared Becard	4	1,320	1,630	1,590	R/Pb
Pachyramphus aglaiae	Rose-throated Becard	31	125	1,580	400	C/Sb
Tityra semifasciata	Masked Tityra	1	600	—	—	A/?
LANIIDAE						
Lanius ludovicianus	Loggerhead Shrike	13	155	1,630	400	U/?b U/W
VIREONIDAE						
Vireo griseus	White-eyed Vireo	1	350	—	—	A/M
Vireo bellii	Bell's Vireo	15	100	600	270	C/Sbp C/W
Vireo atricapillus	Black-capped Vireo	1	260	—	—	A/M
Vireo vicinior	Gray Vireo	5	155	390	260	R/W

Scientific Name[a]	Common Name[a]	N[b]	Min. El.[c]	Max. El.[d]	Med. El.[e]	Abun./ Stat.[f]
Vireo solitarius[g]	Solitary Vireo	26	125	1,600	470	C/Sb C/W
Vireo huttoni	Hutton's Vireo	8	260	1,630	1,500	U/Sb U/W
Vireo hypochryseus	Golden Vireo	2	600	620	—	A/?
Vireo gilvus	Warbling Vireo	21	125	1,630	440	C/Sb C/W
Vireo flavoviridis	Yellow-green Vireo	8	260	1,420	540	U/Sbp
CORVIDAE						
Cyanocitta stelleri	Steller's Jay	4	390	1,600	1,380	U/Pbp
Calocitta colliei	Black-throated Magpie-Jay	41	100	1,420	350	C/Pb
Cyanocorax beecheii	Purplish-backed Jay	19	125	910	340	U/Pb
Aphelocoma ultramarina	Mexican Jay	10	880	1,630	1,490	C/Pb
Corvus sinaloae	Sinaloa Crow	22	125	1,000	275	C/Pb
Corvus corax	Common Raven	36	125	1,630	365	C/Pb
HIRUNDINIDAE						
Progne sinaloae	Sinaloa Martin	2	1,500	1,600	—	R/S
Tachycineta bicolor	Tree Swallow	1	390	—	—	C/M
Tachycineta albilinea	Mangrove Swallow	1	155	—	—	U/?
Tachycineta thalassina	Violet-green Swallow	11	155	1,580	460	U/W
Stelgidopteryx serripennis	Northern Rough-winged Swallow	17	135	1,630	350	U/Sb U/W
Petrochelidon pyrrhonota	Cliff Swallow	7	135	480	260	U/Sb
Hirundo rustica	Barn Swallow	11	100	1,630	435	C/Sb C/M
PARIDAE						
Poecile sclateri	Mexican Chickadee	1	1,380	—	—	R/?
Baeolophus wollweberi	Bridled Titmouse	11	620	1,630	1,480	C/Pb
REMIZIDAE						
Auriparus flaviceps	Verdin	21	100	480	240	C/Pb
AEGITHALIDAE						
Psaltriparus minimus	Bushtit	7	1,380	1,630	1,500	U/Pbp
SITTIDAE						
Sitta canadensis	Red-breasted Nuthatch	2	390	400	—	R/W
Sitta carolinensis	White-breasted Nuthatch	9	240	1,630	1,500	U/Pbp

Scientific Name[a]	Common Name[a]	N^b	Min. El.[c]	Max. El.[d]	Med. El.[e]	Abun./ Stat.[f]
Sitta pygmaea	Pygmy Nuthatch	3	400	1,500	1,380	R/Pbp
CERTHIIDAE						
Certhia americana	Brown Creeper	9	880	1,630	1,480	U/Pbp
TROGLODYTIDAE						
Campylorhynchus gularis	Spotted Wren	14	400	1,630	1,400	U/Pb
Campylorhynchus brunneicapillus	Cactus Wren	11	125	1,380	220	C/Pb
Salpinctes obsoletus	Rock Wren	11	155	800	390	C/W
Catherpes mexicanus	Canyon Wren	32	125	1,630	480	C/Pb
Thryothorus sinaloa	Sinaloa Wren	35	125	1,630	420	C/Pb
Thryothorus felix	Happy Wren	20	100	1,000	295	U/Pb
Thryomanes bewickii	Bewick's Wren	2	240	250	—	R/W
Troglodytes aedon	House Wren	18	155	1,600	415	R/Pbp U/W
REGULIDAE						
Regulus satrapa	Golden-crowned Kinglet	1	420	—	—	A/W
Regulus calendula	Ruby-crowned Kinglet	15	240	1,500	480	C/W
SYLVIIDAE						
Polioptila caerulea	Blue-gray Gnatcatcher	17	150	1,600	580	C/Pb C/W
Polioptila melanura	Black-tailed Gnatcatcher	1	520	—	—	R/Pbp
Polioptila nigriceps	Black-capped Gnatcatcher	37	125	1,500	390	C/Pb
TURDIDAE						
Sialia sialis	Eastern Bluebird	10	380	1,630	1,490	C/Pb
Sialia mexicana	Western Bluebird	2	390	1,380	—	R/W
Sialia currucoides	Mountain Bluebird	1	390	—	—	A/W
Myadestes townsendi	Townsend's Solitaire	2	390	520	—	R/W
Myadestes occidentalis	Brown-backed Solitaire	18	240	1,630	645	U/Sbp U/W
Catharus aurantiirostris	Orange-billed Nightingale-Thrush	6	240	1,600	320	R/Sbp R/W
Catharus ustulatus	Swainson's Thrush	8	260	1,580	550	U/M
Catharus guttatus	Hermit Thrush	17	220	1,630	520	C/W
Turdus assimilis	White-throated Robin	12	390	1,600	940	U/Pb
Turdus rufopalliatus	Rufous-backed Robin	25	155	800	400	C/Pb

Scientific Name[a]	Common Name[a]	N[b]	Min. El.[c]	Max. El.[d]	Med. El.[e]	Abun./ Stat.[f]
Turdus migratorius	American Robin	13	240	1,630	1,380	C/Sb U/W
MIMIDAE						
Dumetella carolinensis	Gray Catbird	1	250	—	—	A/W
Mimus polyglottos	Northern Mockingbird	23	150	1,630	390	C/Sb C/W
Toxostoma bendirei	Bendire's Thrasher	2	240	390	—	R/W
Toxostoma curvirostre	Curve-billed Thrasher	33	100	1,630	350	C/Pb
Melanotis caerulescens	Blue Mocking-bird	26	210	1,630	480	U/Pb
STURNIDAE						
Sturnus vulgaris	European Starling	1	155	—	—	U/P
MOTACILLIDAE						
Anthus rubescens	American Pipit	3	250	390	260	R/W
BOMBYCILLIDAE						
Bombycilla cedrorum	Cedar Waxwing	5	250	620	400	U/W
Ptilogonys cinereus	Gray Silky-flycatcher	5	390	1,630	1,500	R/Sb
Phainopepla nitens	Phainopepla	8	150	440	275	U/?
PEUCEDRAMIDAE						
Peucedramus taeniatus	Olive Warbler	2	1,380	1,420	—	R/Pbp
PARULIDAE						
Vermivora peregrina	Tennessee Warbler	1	520	—	—	A/M
Vermivora celata	Orange-crowned Warbler	23	100	1,380	380	C/W
Vermivora ruficapilla	Nashville Warbler	15	220	670	380	C/W
Vermivora virginiae	Virginia's Warbler	1	400	—	—	A/M
Vermivora luciae	Lucy's Warbler	13	135	800	350	R/Sbp C/M
Parula superciliosa	Crescent-chested Warbler	1	1,380	—	—	R/S
Parula americana	Northern Parula	1	420	—	—	A/M
Parula pitiayumi	Tropical Parula	9	260	1,420	580	U/Sb
Dendroica petechia	Yellow Warbler	11	155	1,580	400	R/Sb C/M
Dendroica magnolia	Magnolia Warbler	2	320	390	—	A/W
Dendroica coronata	Yellow-rumped Warbler	19	220	1,500	390	C/W

Scientific Name[a]	Common Name[a]	N[b]	Min. El.[c]	Max. El.[d]	Med. El.[e]	Abun./ Stat.[f]
Dendroica nigrescens	Black-throated Gray Warbler	20	155	1,500	415	C/W
Dendroica townsendi	Townsend's Warbler	10	320	1,580	735	U/M R/W
Dendroica occidentalis	Hermit Warbler	4	390	1,500	1,380	U/M R/W
Dendroica graciae	Grace's Warbler	9	880	1,630	1,480	C/Sb
Mniotilta varia	Black-and-white Warbler	4	240	1,380	455	U/W
Protonotaria citrea	Prothonotary Warbler	1	260	—	—	A/M
Seiurus noveboracensis	Northern Waterthrush	3	260	670	390	U/W
Seiurus motacilla	Louisiana Waterthrush	5	240	400	320	R/W
Oporornis formosus	Kentucky Warbler	1	1,380	—	—	A/M
Oporornis tolmiei	MacGillivray's Warbler	23	155	1,580	400	C/W
Geothlypis trichas	Common Yellowthroat	12	100	620	310	U/Sb U/W
Wilsonia citrina	Hooded Warbler	1	260	—	—	A/W
Wilsonia pusilla	Wilson's Warbler	19	155	1,580	390	C/W
Myioborus pictus	Painted Redstart	18	260	1,630	1,100	C/W
Myioborus miniatus	Slate-throated Redstart	8	260	1,500	500	U/Sbp
Euthlypis lachrymosa	Fan-tailed Warbler	4	400	1,500	750	R/Sbp
Basileuterus rufifrons	Rufous-capped Warbler	18	400	1,630	645	U/Pb
Icteria virens	Yellow-breasted Chat	13	155	620	270	C/Sb
THRAUPIDAE						
Piranga flava	Hepatic Tanager	16	250	1,630	1,380	C/Sb C/W
Piranga rubra	Summer Tanager	15	125	1,420	265	C/Sb U/W
Piranga ludoviciana	Western Tanager	20	155	1,630	475	C/W
Piranga bidentata	Flame-colored Tanager	5	390	1,320	670	R/?
Piranga erythrocephala	Red-headed Tanager	1	620	—	—	A/?S
Euphonia affinis	Scrub Euphonia	2	390	800	—	R/?P
Euphonia elegantissima	Blue-hooded Euphonia	6	390	1,420	1,090	R/?S

Scientific Name[a]	Common Name[a]	N[b]	Min. El.[c]	Max. El.[d]	Med. El.[e]	Abun./ Stat.[f]
EMBERIZIDAE						
Volatinia jacarina	Blue-black Grassquit	5	135	880	205	R/Sbp
Melozone kieneri	Rusty-crowned Ground-Sparrow	5	400	1,000	600	R/?
Pipilo chlorurus	Green-tailed Towhee	17	150	1,380	350	C/W
Pipilo maculatus	Spotted Towhee	1	800	—	—	A/W
Pipilo fuscus	Canyon Towhee	22	125	1,630	310	C/Pb
Aimophila carpalis	Rufous-winged Sparrow	13	125	400	240	C/Pb
Aimophila cassinii	Cassin's Sparrow	3	155	400	390	R/W
Aimophila botterii	Botteri's Sparrow	5	270	1,380	480	R/W
Aimophila ruficeps	Rufous-crowned Sparrow	12	275	1,630	1,450	C/Pb
Aimophila rufescens	Rusty Sparrow	10	620	1,630	1,365	U/Pb
Aimophila quinquestriata	Five-striped Sparrow	12	125	910	480	C/Pbp
Spizella passerina	Chipping Sparrow	22	155	1,600	460	C/Sb C/W
Spizella pallida	Clay-colored Sparrow	6	240	1,380	325	U/W
Spizella breweri	Brewer's Sparrow	6	150	380	255	U/W
Spizella atrogularis	Black-chinned Sparrow	1	800	—	—	A/W
Pooecetes gramineus	Vesper Sparrow	7	155	1,380	350	U/W
Chondestes grammacus	Lark Sparrow	12	150	1,380	315	C/W
Amphispiza bilineata	Black-throated Sparrow	3	155	390	240	R/W
Calamospiza melanocorys	Lark Bunting	3	150	910	260	U/W
Passerculus sandwichensis	Savannah Sparrow	2	260	350	—	U/W
Ammodramus savannarum	Grasshopper Sparrow	4	155	400	315	U/W
Passerella iliaca	Fox Sparrow	1	240	—	—	A/W
Melospiza melodia	Song Sparrow	2	220	275	—	R/W
Melospiza lincolnii	Lincoln's Sparrow	16	100	1,380	320	C/W
Zonotrichia albicollis	White-throated Sparrow	1	390	—	—	A/W

Scientific Name[a]	Common Name[a]	N^b	Min. El.[c]	Max. El.[d]	Med. El.[e]	Abun./ Stat.[f]
Zonotrichia leucophrys	White-crowned Sparrow	14	155	1,380	365	C/W
Zonotrichia atricapilla	Golden-crowned Sparrow	1	390	—	—	A/W
Junco hyemalis	Dark-eyed Junco	4	1,380	1,500	1,400	U/W
Junco phaeonotus	Yellow-eyed Junco	1	1,380	—	—	A/?
CARDINALIDAE						
Cardinalis cardinalis	Northern Cardinal	36	125	1,380	350	C/Pb
Cardinalis sinuatus	Pyrrhuloxia	19	100	1,380	320	U/Sb C/W
Pheucticus chrysopeplus	Yellow Grosbeak	20	125	1,350	370	C/Sb
Pheucticus ludovicianus	Rose-breasted Grosbeak	1	800	—	—	A/M
Pheucticus melanocephalus	Black-headed Grosbeak	29	125	1,630	400	U/Sb C/W
Guiraca caerulea	Blue Grosbeak	16	135	1,600	440	C/Sbp
Passerina amoena	Lazuli Bunting	18	125	1,380	335	C/W
Passerina cyanea	Indigo Bunting	4	250	390	320	R/W
Passerina versicolor	Varied Bunting	32	125	800	365	C/Sb C/W
Passerina ciris	Painted Bunting	5	100	480	260	U/M
ICTERIDAE						
Agelaius phoeniceus	Red-winged Blackbird	2	160	390	—	U/Pbp
Sturnella magna	Eastern Meadowlark	1	390	—	—	A/W
Sturnella neglecta	Western Meadowlark	2	155	390	—	U/W
Xanthocephalus xanthocephalus	Yellow-headed Blackbird	2	260	400	—	A/W
Euphagus cyanocephalus	Brewer's Blackbird	1	390	—	—	A/W
Quiscalus mexicanus	Great-tailed Grackle	11	155	520	220	C/Pb
Molothrus aeneus	Bronzed Cowbird	20	125	1,630	395	C/Sb
Molothrus ater	Brown-headed Cowbird	19	125	1,630	390	C/Pb
Icterus wagleri	Black-vented Oriole	20	205	1,630	550	C/Pb
Icterus spurius	Orchard Oriole	5	100	390	240	U/Sbp
Icterus cucullatus	Hooded Oriole	13	160	520	320	C/Pb

Scientific Name[a]	Common Name[a]	N[b]	Min. El.[c]	Max. El.[d]	Med. El.[e]	Abun./ Stat.[f]
Icterus pustulatus	Streak-backed Oriole	37	125	1,630	380	C/Pb
Icterus galbula	Baltimore Oriole	1	390	—	—	A/M
Icterus bullockii	Bullock's Oriole	10	220	800	390	U/M R/W
Icterus parisorum	Scott's Oriole	9	800	1,630	1,420	U/Sb
Cacicus melanicterus	Yellow-winged Cacique	1	260	—	—	A/?
FRINGILLIDAE						
Carpodacus purpureus	Purple Finch	2	250	260	—	?/W
Carpodacus mexicanus	House Finch	23	125	1,630	275	C/Pb
Loxia curvirostra	Red Crossbill	1	1,500	—	—	A/W
Carduelis pinus	Pine Siskin	2	1,380	1,420	—	R/W
Carduelis notata	Black-headed Siskin	8	880	1,630	1,460	U/Pb
Carduelis psaltria	Lesser Goldfinch	23	125	1,630	400	C/Pbp
Carduelis lawrencei	Lawrence's Goldfinch	1	390	—	—	A/W
Coccothraustes abeillei	Hooded Grosbeak	2	1,320	1,500	—	A/S
Coccothraustes vespertinus	Evening Grosbeak	1	1,380	—	—	A/M
PASSERIDAE						
Passer domesticus	House Sparrow	6	160	1,500	300	C/Pb

Source: Russell and Monson (1998). Species listed in this table are those recorded at elevations of 100 m or greater in an area bounded on the south by the Sinaloa border, on the east by the Chihuahua border, on the north by 27°45'N latitude, and on the west by 109°15'W longitude (north of 26°45'N) and 109°00'W longitude (south of 26°45'N).

[a] Scientific and common names follow the *Check-list of North American Birds* (AOU 1998).

[b] N = the number of localities (68 possible) in which the species was found in the Alamos area.

[c] Min. El. = the lowest elevation (m) at which the species was found.

[3] Max. El. = the highest elevation (m) at which the species was found.

[e] Med. El. = the median elevation (m) at which the species was found.

[f] Abun./Stat. = abundance/status. Abundance: C = common (individuals, pairs, or groups frequently encountered); U = uncommon (the species is present but is not always found); R = rare (the species is present but seldom found); A = accidental (five or fewer records). Status (seasonal distribution): P = permanent resident (present year-round); S = summer resident that breeds in the area but is absent much of the year (typically present from April to October); M = migrant (moves through the area in spring and/or autumn); W = winter migrant that is present December into February and as a migrant at other times; b = breeding confirmed; bp = breeding presumed.

[g] Includes *V. cassinii* (Cassin's Vireo) and *V. plumbeus* (Plumbeous Vireo).

Scientific Name[a]	Common Name[a]	H[b]	A[c]	S[d]	C[e]	Breeding Range[f]
CICONIIFORMES						
ARDEIDAE						
Tigrisoma mexicanum	Bare-throated Tiger-Heron Garza-tigre Mexicana	R	U	Pb		s Sonora to Colombia
CATHARTIDAE						
Coragyps atratus	Black Vulture Zopilote Común	FRSO	C	Pbp		s Arizona to Argentina
Cathartes aura	Turkey Vulture Zopilote Aura	FRSO	C	Pbp		Canada to Argentina
FALCONIFORMES						
ACCIPITRIDAE						
Elanus leucurus	White-tailed Kite Milano Cola Blanca	SO	R	Pbp	1, 2	Oregon to Argentina
Accipiter striatus	Sharp-shinned Hawk Gavilán Pecho Rufo	FRS	C	W		Alaska to s Mexico
Accipiter cooperii	Cooper's Hawk Gavilán de Cooper	FRS	C	W		Canada to n Mexico
Geranospiza caerulescens	Crane Hawk Gavilán Zancón	FRS	R	Pbp		s Sonora to Argentina
Asturina nitidus	Gray Hawk Aguililla Gris	FRS	C	Pb		s Arizona to Argentina
Buteogallus anthracinus	Common Black-Hawk Aguililla-negra Menor	FR	U	Pb		Arizona to Venezuela
Buteogallus urubitinga	Great Black-Hawk Aguililla-negra Major	R	R	?		s Sonora to Argentina
Harpyhaliaetus solitarius	Solitary Eagle Águila Solitaria	FR	R	?b	3	s Sonora to Argentina
Buteo brachyurus	Short-tailed Hawk Aguililla Cola Corta	F	R	P?	1	s Sonora, Florida to Argentina

Scientific Name[a]	Common Name[a]	H[b]	A[c]	S[d]	C[e]	Breeding Range[f]
Buteo albicaudatus	White-tailed Hawk Aguililla Cola Blanca	SO	R	P?	3	s Texas to Argentina
Buteo albonotatus	Zone-tailed Hawk Aguililla Aura	FRS	U	Sb		Arizona to Brazil
Buteo jamaicensis	Red-tailed Hawk Aguililla Cola Roja	SO	C	W		Alaska to Panama
FALCONIDAE						
Caracara plancus	Crested Caracara Caracara Quebrantahuesos	FRSO	C	Pb		s Arizona, Florida to Argentina
Herpetotheres cachinnans	Laughing Falcon Halcón Guaco	FS	R	Pb	3	s Sonora to Argentina
Falco sparverius	American Kestrel Cernícalo Americano	FRSO	C	W		Alaska to Argentina
Falco rufigularis	Bat Falcon Halcón Enano	SO	R	Pb	3	s Sonora to Argentina
GALLIFORMES						
CRACIDAE						
Ortalis wagleri	Rufous-bellied Chachalaca Chachalaca Vientre Castaño	FS	U	Pb		s Sonora to n Jalisco
PHASIANIDAE						
Callipepla douglasii	Elegant Quail Codorniz Cresta Dorada	FS	C	Pbp		n Sonora to n Jalisco
COLUMBIFORMES						
COLUMBIDAE						
Columba flavirostris	Red-billed Pigeon Paloma Morada	FRS	C	Pb		s Sonora to Costa Rica
Zenaida asiatica	White-winged Dove Paloma Ala Blanca	FRSO	C	Pb		California to Chile

Appendix 7.2. (Continued)

Scientific Name[a]	Common Name[a]	H[b]	A[c]	S[d]	C[e]	Breeding Range[f]
Zenaida macroura	Mourning Dove	FRSO	C	PbW		Canada to Panama
Columbina inca	Paloma Huilota Inca Dove Tórtola Cola Larga	SRO	C	Pb		California to Costa Rica
Columbina passerina	Common Ground-Dove Tórtola Coquita	FRSO	C	Pb		California to Brazil
Columbina talpacoti	Ruddy Ground-Dove Tórtola Rojiza	SO	R	Pbp		s Sonora to Argentina
Leptotila verreauxi	White-tipped Dove Paloma Arroyera	FRS	C	Pb		c Sonora to Argentina
PSITTACIFORMES PSITTACIDAE						
Ara militaris	Military Macaw Guacamaya Verde	F	R	Sb		s Sonora to Argentina
Forpus cyanopygius	Mexican Parrotlet Perico Catarina	FS	R	Sbp	3	s Sonora to Jalisco
Amazona albifrons	White-fronted Parrot Loro Frente Blanca	FRS	C	Pb		s Sonora to nw Costa Rica
Amazona finschi	Lilac-crowned Parrot Loro Corona Lila	F	R	Pb		s Sonora to Oaxaca
CUCULIFORMES CUCULIDAE						
Coccyzus americanus	Yellow-billed Cuckoo Cuclillo Pico Amarillo	FRS	C	Sb		s Canada to n Mexico
Coccyzus minor	Mangrove Cuckoo Cuclillo Manglero	FS	U	Sb	1, 2	s Sonora to Brazil

Scientific Name[a]	Common Name[a]	H[b]	A[c]	S[d]	C[e]	Breeding Range[f]
Piaya cayana	Squirrel Cuckoo Cuclillo Canela	FR	U	Pb		s Sonora to Argentina
Geococcyx velox	Lesser Road-runner Correcaminos Tropical	F	R	Pb		s Sonora to Oaxaca
Geococcyx californianus	Greater Roadrunner Correcaminos Norteño	SO	C	Pb		California, Texas to c Mexico
Crotophaga sulcirostris	Groove-billed Ani Garrapatero Pijuy	SRO	U	Sb		s Sonora to Argentina
STRIGIFORMES						
STRIGIDAE						
Otus kennicottii	Western Screech-Owl Tecolote Occidental	FRS	C	Pb		Alaska to c Mexico
Otus guatemalae	Vermiculated Screech-Owl Tecolote Vermiculado	F	R	Pb		s Sonora to Bolivia
Bubo virginianus	Great Horned Owl Búho Cornudo	FRS	C	Pbp		Alaska to Argentina
Glaucidium griseiceps	Central American Pygmy-Owl Tecolote Colimense	F	R	Pb		s Sonora to Brazil
Glaucidium brasilianum	Ferruginous Pygmy-Owl Tecolote Bajeño	FRS	U	Pb		s Arizona to Argentina
Micrathene whitneyi	Elf Owl Tecolote Enano	FRS	C	Sbp		s Nevada to c Mexico
Ciccaba virgata	Mottled Owl Búho Café	FR	R	Pb		s Sonora to Argentina

Appendix 7.2. (Continued)

Scientific Name[a]	Common Name[a]	H[b]	A[c]	S[d]	C[e]	Breeding Range[f]
CAPRIMULGIFORMES						
CAPRIMULGIDAE						
Chordeiles acutipennis	Lesser Night-hawk Chotacabras Menor	so	C	Sbp		California to Brazil
Nyctidromus albicollis	Common Pauraque Chotacabras Pauraque	so	R	Sb		s Sonora to Argentina
Caprimulgus ridgwayi	Buff-collared Nightjar Tapacamino Tu-cuchillo	FRS	C	Sb		s Arizona to Nicaragua
APODIFORMES						
APODIDAE						
Aeronautes saxatalis	White-throated Swift Vencejo Pecho Blanco	A	U	Pbp		s Canada to Honduras
TROCHILIDAE						
Cynanthus latirostris	Broad-billed Hummingbird Colibrí Pico Ancho	FRS	C	Pb		s Arizona to Chiapas
Amazilia beryllina	Berylline Hummingbird Colibrí Berilo	FRS	R	W		c Sonora to Honduras
Amazilia violiceps	Violet-crowned Hummingbird Colibrí Corona Violeta	FR	C	Pb		s Arizona to Oaxaca
Heliomaster constantii	Plain-capped Starthroat Colibrí Picudo	F	U	Sb		s Sonora to Costa Rica
Calypte costae	Costa's Hummingbird Colibrí Cabeza Violeta	F	U	W		s Utah to c Sonora
Selasphorus rufus	Rufous Hummingbird Zumbador Rufo	FRS	U	M		s Alaska to Oregon, Idaho

Scientific Name[a]	Common Name[a]	H[b]	A[c]	S[d]	C[e]	Breeding Range[f]
TROGONIFORMES						
TROGONIDAE						
Trogon elegans	Elegant Trogon Trogón Elegante	FRS	C	Pb		s Arizona to nw Costa Rica
CORACIIFORMES						
MOMOTIDAE						
Momotus mexicanus	Russet-crowned Motmot Momoto Corona Café	F	U	Sb		s Sonora to Guatemala
PICIFORMES						
PICIDAE						
Melanerpes uropygialis	Gila Woodpecker Carpintero del Desierto	FRS	C	Pb		s Nevada to Jalisco
Picoides scalaris	Ladder-backed Woodpecker Carpintero Mexicano	FS	C	Pb		California to Nicaragua
Colaptes chrysoides	Gilded Flicker Carpintero Collarejo Desértico	FS	U	Pb		s California to n Sinaloa
Dryocopus lineatus	Lineated Woodpecker Carpintero Lineado	F	R	Pbp	3	s Sonora to Argentina
Campephilus guatemalensis	Pale-billed Woodpecker Carpintero Pico Plata	F	R	Pb	3	s Sonora to Panama
PASSERIFORMES						
DENDROCOLAPTIDAE						
Xiphlorhynchus flavigaster	Ivory-billed Woodcreeper Trepatroncos Bigotudo	F	R	Pb	3	s Sonora, Tamaulipas to Costa Rica
Lepidocolaptes leucogaster	White-striped Woodcreeper Trepatroncos Escarchado	F	U	Sbp		s Sonora to Oaxaca

Appendix 7.2. (Continued)

Scientific Name[a]	Common Name[a]	H[b]	A[c]	S[d]	C[e]	Breeding Range[f]
TYRANNIDAE						
Camptostoma imberbe	Northern Beardless-Tyrannulet Mosquero Lampiño	FRS	C	Pb		s Arizona to nw Costa Rica
Mitrephanes phaeocercus	Tufted Flycatcher Mosquero Copetón	R	U	W		Sonora to Bolivia
Contopus pertinax	Greater Pewee Pibí Tengofrío	FRS	U	W		c Arizona to Nicaragua
Contopus sordidulus	Western Wood-Pewee Pibí Occidental	FRS	C	M		Alaska to Honduras
Empidonax traillii	Willow Flycatcher Mosquero Saucero	FRS	U	M		Canada to California, Texas
Empidonax hammondii	Hammond's Flycatcher Mosquero de Hammond	FRS	U	M		Alaska to California, New Mexico
Empidonax wrightii	Gray Flycatcher Mosquero Gris	FRS	C	W		Washington to California, New Mexico
Empidonax oberholseri	Dusky Flycatcher Mosquero Oscuro	FRS	C	W		Canada to s California, Colorado
Empidonax difficilis	Pacific-slope Flycatcher Mosquero Californiano	FRS	C	W		Alaska to Baja California
Empidonax fulvifrons	Buff-breasted Flycatcher Mosquero Pecho Leonado	FRS	U	W		Arizona to Honduras
Pyrocephalus rubinus	Vermilion Flycatcher Mosquero Cardenal	SO	C	Pb		s California to Argentina

Scientific Name[a]	Common Name[a]	H[b]	A[c]	S[d]	C[e]	Breeding Range[f]
Attila spadiceus	Bright-rumped Attila / Atila	F	R	?Pbp	I	s Sonora to Brazil
Myiarchus tuberculifer	Dusky-capped Flycatcher / Papamoscas Triste	FRS	C	?Pb W		Arizona to Argentina
Myiarchus cinerascens	Ash-throated Flycatcher / Papamoscas Cenizo	FS	C	W		Washington to Jalisco
Myiarchus nuttingi	Nutting's Flycatcher / Papamoscas de Nutting	FS	C	Pb		c Sonora to nw Costa Rica
Myiarchus tyrannulus	Brown-crested Flycatcher / Papamoscas Tirano	FR	U	Sb		sw Utah to n Argentina
Myiodynastes luteiventris	Sulphur-bellied Flycatcher / Papamoscas Atigrado	FR	C	Sb		s Arizona to c Costa Rica
Tyrannus melancholicus	Tropical Kingbird / Tirano Tropical	SR	C	Sb		s Arizona to Argentina
Tyrannus vociferans	Cassin's Kingbird / Tirano Gritón	SRO	U	W		Utah to Oaxaca
Tyrannus crassirostris	Thick-billed Kingbird / Tirano Pico Grueso	FRS	C	Sb		s Arizona to Oaxaca
Tyrannus verticalis	Western Kingbird / Tirano Pálido	SRO	U	M		s Canada to n Sonora
Pachyramphus aglaiae	Rose-throated Becard / Mosquero-cabezón Degollado	FR	C	Sb		s Arizona to nw Costa Rica
Tityra semifasciata	Masked Tityra / Titira Enmascarada	F	A	?	3	s Sonora to Brazil

Scientific Name[a]	Common Name[a]	H[b]	A[c]	S[d]	C[e]	Breeding Range[f]
VIREONIDAE						
Vireo bellii	Bell's Vireo Vireo de Bell	SR	C	W		California to Zacatecas
Vireo vicinior	Gray Vireo Vireo Gris	SR	R	W		s Utah to Texas
Vireo solitarius	Solitary Vireo Vireo Anteojillo	FRS	C	W		Canada to Honduras
Vireo hypochryseus	Golden Vireo Vireo Dorado	FR	A	?P	3	s Sonora to Oaxaca
Vireo gilvus	Warbling Vireo Vireo Gojeador	FRS	C	W		Alaska to Oaxaca
Vireo flavoviridis	Yellow-green Vireo Vireo Verde- amarillo	FR	U	Sbp		s Sonora to Panama
CORVIDAE						
Calocitta colliei	Black-throated Magpie-Jay Urraca-hermosa Cara Negra	FS	C	Pb		s Sonora to Jalisco
Cyanocorax beecheii	Purplish-backed Jay Chara de Beechy	F	U	Pb		s Sonora to Nayarit
Corvus sinaloae	Sinaloa Crow Cuervo Sinaloense	FRSO	C	Pb		s Sonora to Colima
Corvus corax	Common Raven Cuervo Común	FRSO	C	Pb		Alaska to Nicaragua
TROGLODYTIDAE						
Campylorhynchus gularis	Spotted Wren Matraca Desértica	F	U	Pbp		s Sonora to c Mexico
Salpinctes obsoletus	Rock Wren Chivirín Saltarroca	FS	U	W		s Canada to nw Costa Rica
Catherpes mexicanus	Canyon Wren Chivirín Barranqueño	FRS	U	Pb		s Canada to Chiapas
Thryothorus sinaloa	Sinaloa Wren Chivirín Sinaloense	FRS	C	Pb		n Sonora to Oaxaca
Thryothorus felix	Happy Wren Chivirín Feliz	FRS	U	Pb		s Sonora to Oaxaca

Scientific Name[a]	Common Name[a]	H[b]	A[c]	S[d]	C[e]	Breeding Range[f]
Troglodytes aedon	House Wren Chivirín Saltapared	FRS	C	W		Canada to Argentina
REGULIDAE						
Regulus calendula	Ruby-crowned Kinglet Reyezuelo de Rojo	FRS	C	W		Alaska to s Arizona
SYLVIIDAE						
Polioptila caerulea	Blue-gray Gnatcatcher Perlita Azulgris	FRS	C	W		s Oregon to Chiapas
Polioptila nigriceps	Black-capped Gnatcatcher Perlita Sinaloense	FRS	C	Pb		s Arizona to Colima
TURDIDAE						
Catharus aurantiirostris	Orange-billed Nightingale-Thrush Zorzal Pico Naranja	R	R	W		sw Chihuahua to Venezuela
Catharus guttatus	Hermit Thrush Zorzal Cola Rufa	FRS	C	W		Alaska to s Arizona
Turdus assimilis	White-throated Robin Mirlo Garganta Blanca	R	U	PbW		s Sonora to Ecuador
Turdus rufopalliatus	Rufous-backed Robin Mirlo Dorso Rufo	FR	C	Pb		s Sonora to Oaxaca
MIMIDAE						
Mimus polyglottos	Northern Mockingbird Centzontle Norteño	SO	C	W		Oregon to Oaxaca
Toxostoma curvirostre	Curve-billed Thrasher Cuitacoche Pico Curvo	FRSO	C	Pb		Colorado to Oaxaca
Melanotis caerulescens	Blue Mockingbird Mulato Azul	FR	U	Pb		s Sonora to Oaxaca

Scientific Name[a]	Common Name[a]	H[b]	A[c]	S[d]	C[e]	Breeding Range[f]
PARULIDAE						
Vermivora celata	Orange-crowned Warbler Chipe Corona Naranja	FRS	C	W		Alaska to w Texas
Vermivora ruficapilla	Nashville Warbler Chipe de Coronilla	FRS	C	W		Idaho to w Texas
Vermivora luciae	Lucy's Warbler Chipe Rabadilla Rufa	FRS	C	M		Utah to n Sonora
Parula pitiayumi	Tropical Parula Parula Tropical	FR	U	Sb		Sonora, Texas to Argentina
Dendroica petechia	Yellow Warbler Chipe Amarillo	FRS	C	M		Alaska to Peru
Dendroica coronata	Yellow-rumped Warbler Chipe Coronado	FRS	C	W		Alaska to Guatemala
Dendroica nigrescens	Black-throated Gray Warbler Chipe Negrogris	FRS	C	W		sw Canada to n Sonora
Oporornis tolmiei	MacGillivray's Warbler Chipe de Tolmie	FRS	C	W		Alaska to s New Mexico
Wilsonia pusilla	Wilson's Warbler Chipe Corona Negra	FRS	C	W		Alaska to New Mexico
Myioborus pictus	Painted Redstart Chipe Ala Blanca	FRS	C	W		Arizona to Nicaragua
Basileuterus rufifrons	Rufous-capped Warbler Chipe Gorra Rufa	FRS	U	Pb		n Sonora to Venezuela
Icteria virens	Yellow-breasted Chat Buscabreña	SR	C	Sb		s Canada to Jalisco
THRAUPIDAE						
Piranga flava	Hepatic Tanager Tángara Encinera	FRS	C	W		California to Nicaragua

Scientific Name[a]	Common Name[a]	H[b]	A[c]	S[d]	C[e]	Breeding Range[f]
Piranga rubra	Summer Tanager Tángara Roja	FRS	U	W		Utah to n Mexico
Piranga ludoviciana	Western Tanager Tángara Capucha Roja	FRS	C	W		Alaska to w Texas
Euphonia affinis	Scrub Euphonia Eufonia Garganta Negra	FS	R	?P		s Sonora to nw Costa Rica
EMBERIZIDAE						
Volatinia jacarina	Blue-black Grassquit Semillero Brincador	O	R	Sbp		s Sonora to Argentina
Pipilo chlorurus	Green-tailed Towhee Toquí Cola Verde	FRS	C	W		Washington to w Texas
Pipilo fuscus	Canyon Towhee Toquí Pardo	FRS	C	Pb		Oregon to Oaxaca
Aimophila quinquestriata	Five-striped Sparrow Zacatonero Cinco Rayas	FS	C	Pbp		s Arizona to Jalisco
Spizella passerina	Chipping Sparrow Gorrión Ceja Blanca	SO	C	W		Alaska to Nicaragua
Chondestes grammacus	Lark Sparrow Gorrión Arlequín	SO	C	W		s Canada to Zacatecas
Melospiza lincolnii	Lincoln's Sparrow Gorrión de Lincoln	S	C	W		Alaska to New Mexico
Zonotrichia leucophrys	White-crowned Sparrow Gorrión Corona Blanca	SO	C	W		Alaska to New Mexico
CARDINALIDAE						
Cardinalis cardinalis	Northern Cardinal Cardenal Rojo	SR	C	Pb		California to Belize

Scientific Name[a]	Common Name[a]	H[b]	A[c]	S[d]	C[e]	Breeding Range[f]
Cardinalis sinuatus	Pyrrhuloxia Cardenal Pardo	SR	C	W		s Arizona to Jalisco
Pheucticus chrysopeplus	Yellow Grosbeak Picogordo Amarillo	FR	C	Sb		s Sonora to Guatemala
Pheucticus melanocephalus	Black-headed Grosbeak Picogordo Tigrillo	FRS	C	W		s Canada to Oaxaca
Guiraca caerulea	Blue Grosbeak Picogordo Azul	SRO	C	Sbp		Utah to c Costa Rica
Passerina amoena	Lazuli Bunting Colorín Lázuli	SRO	C	W		s Canada to Texas
Passerina versicolor	Varied Bunting Colorín Morado	FRSO	C	SbW		s Arizona to Guatemala
ICTERIDAE						
Molothrus aeneus	Bronzed Cowbird Tordo Ojo Rojo	FRSO	C	Sb		Arizona to Panama
Molothrus ater	Brown-headed Cowbird Tordo Cabeza Café	SRO	C	Pb		Alaska to c Mexico
Icterus wagleri	Black-vented Oriole Bolsero de Wagler	FR	C	Pb		s Sonora to Nicaragua
Icterus cucullatus	Hooded Oriole Bolsero Encapuchado	SRO	U	Pb		s Utah to Belize
Icterus pustulatus	Streak-backed Oriole Bolsero Dorso Rayado	FRS	C	Pb		Sonora to nw Costa Rica
Cacicus melanicterus	Yellow-winged Cacique Cacique Mexicano	F	A	?Pb	3	s Sonora to Chiapas

Scientific Name[a]	Common Name[a]	H[b]	A[c]	S[d]	C[e]	Breeding Range[f]
FRINGILLIDAE						
Carpodacus mexicanus	House Finch	so	C	Pb		Canada to
	Pinzón Mexicano					Oaxaca
Carduelis psaltria	Lesser Goldfinch	sro	C	Pbp W		Washington to
	Jilguero Dominico					Peru

Source: Russell and Monson (1998).

[a] Scientific names and English common names follow the *Check-list of North American Birds* (AOU 1998). Spanish common names follow Escalante et al. (1996).

[b] Habitat (H): A = aerial (forages in air over all habitats); F = tropical deciduous forest; o = open areas, including pastures, cultivated and abandoned fields, and fencerows; R = riparian forest in area of tropical deciduous forest; and s = secondary growth areas.

[c] Abundance (A): C = common (individuals, pairs, or groups frequently encountered); U = uncommon (the species is present but is not always found); R = rare (the species is present but seldom found); A = accidental (five or fewer records).

[d] Status (S; seasonal distribution): P = permanent resident (present year-round); S = summer resident that breeds in the area but is absent much of the year (typically present from April to October); M = migrant (moves through the area in spring and/or autumn); W = winter migrant that is present December into February and as a migrant at other times; b = breeding confirmed; bp = breeding presumed.

[e] Changes in population (C): 1 = not found in the Alamos area before 1950 (may be establishing itself in the area); 2 = numbers increasing since 1950; 3 = numbers have declined or the species has not been found since 1950.

[f] Breeding range: A much condensed summary, from north to south, based on AOU (1998). Compass directions are listed in lowercase letters; c = central.

REFERENCES

Alden, P. 1969. *Finding the birds in western Mexico*. Tucson: University of Arizona Press.

American Ornithologists' Union (AOU). 1998. *Check-list of North American birds*, 7th ed. Washington, D.C.: American Ornithologists' Union.

Arizmendi, M. del C., H. Berlanga, L. Marquez-Vadelamar, L. Navarijo, and F. Ornelas. 1990. *Avifauna de la región de Chamela, Jalisco*. Cuadernos del Instituto de Biología 4. México, D.F.: Universidad Nacional Autónoma de México.

Askins, R. A., J. F. Lynch, and R. Greenberg. 1990. Population declines in migratory birds in eastern North America. *Current Ornithology* 7:1–57.

Escalante, P., A. M. Sada, and J. R. Gil. 1996. *Listado de nombres comunes de las aves de México*. México, D.F.: ASM, Comisíon Nacional para el Conocimiento y Uso de la Biodiversidad.

Harrison, E., and L. Kiff. 1977. The nest and egg of the black solitary eagle. *Condor* 79:132–133.

Hutto, R. L. 1992. Habitat distributions of migratory landbird species in western Mexico. In *Ecology and conservation of Neotropical migrant landbirds*, eds. J. M. Hagan III and D. W. Johnston, 221–239. Washington, D.C.: Smithsonian Institution Press.

Kattan, G. H., H. Alvarez-López, and M. Giraldo. 1994. Forest fragmentations and bird extinctions: San Antonio eighty years later. *Conservation Biology* 8:138–146.

Paton, P.W.C. 1994. The effect of edge on avian nest success: How strong is the evidence? *Conservation Biology* 8:17–26.

Robbins, C. S., D. K. Dawson, and B. A. Dowell. 1989. Habitat area requirements of breeding forest birds of the Middle Atlantic states. *Wildlife Monographs* 103:1–34.

Robinson, S. K., and D. S. Wilcove. 1989. Conserving tropical raptors and game birds. *Conservation Biology* 3:192–193.

Russell, S. M., and G. Monson. 1998. *The birds of Sonora*. Tucson: University of Arizona Press.

Short, L. L. 1974. Nesting of southern Sonora birds during the summer rainy season. *Condor* 76:21–32.

Stiles, F. G. 1983. Birds. In *Costa Rican natural history*, ed. D. H. Janzen, 502–544. Chicago: University of Chicago Press.

Thiollay, J. M. 1989. Area requirements for the conservation of rain forest raptors and game birds in French Guiana. *Conservation Biology* 3:128–137.

van Rossem, A. J. 1945. *A distributional survey of the birds of Sonora, Mexico*. Occasional Papers of the Museum of Zoology No. 21. Baton Rouge: Louisiana State University Press.

Willson, M. F., T. L. De Santo, C. Sabag, and J. J. Armesto. 1994. Avian communities of fragmented south-temperate rainforests in Chile. *Conservation Biology* 8:508–520.

CONTRIBUTORS

BARNEY T. BURNS, Native Seeds/SEARCH, 526 N. Fourth Ave., Tucson, Arizona 85705

ALBERTO BÚRQUEZ, Instituto de Ecología, Universidad Nacional Autónoma de México, Apartado Postal 1354, Hermosillo, CP 83000, Sonora

MAHINA DREES, Native Seeds/SEARCH, 526 N. Fourth Ave., Tucson, Arizona 85705

RIGOBERTO A. LÓPEZ ESTUDILLO, DICTUS, Universidad de Sonora, Apartado Postal 1819, Hermosillo, CP 83000, Sonora

CHARLES H. LOWE, Department of Ecology and Evolutionary Biology, University of Arizona, Tucson, Arizona 85721

MANUEL MAASS, Instituto de Ecología, Universidad Nacional Autónoma de México, Apartado Postal 27-3, Xangari, Morelia, CP 58089, Michoacán

PAUL S. MARTIN, Desert Laboratory, Department of Geosciences, University of Arizona, Tucson, Arizona 85721

ANGELINA MARTÍNEZ-YRÍZAR, Instituto de Ecología, Universidad Nacional Autónoma de México, Apartado Postal 1354, Hermosillo, CP 83000, Sonora

STEPHANIE A. MEYER, Apartado Postal 34, Alamos, CP 85760 Sonora

GARY P. NABHAN, Arizona-Sonora Desert Museum, 2021 N. Kinney Rd., Tucson, Arizona 85743

SUZANNE C. NELSON, Native Seeds/SEARCH, 526 N. Fourth Ave., Tucson, Arizona 85705

ANA LILIA REINA GUERRERO, Arizona-Sonora Desert Museum, 2021 N. Kinney Rd., Tucson, Arizona 85743

STEPHEN M. RUSSELL, Department of Ecology and Evolutionary Biology, University of Arizona, Tucson, Arizona 85721

ANDREW C. SANDERS, Herbarium, Department of Botany and Plant Sciences, University of California, Riverside, California 92521

CECIL R. SCHWALBE, U.S. Geological Survey, Sonoran Desert Field Station, University of Arizona, Tucson, Arizona 85721

THOMAS R. VAN DEVENDER, Arizona-Sonora Desert Museum, 2021 N. Kinney Rd., Tucson, Arizona 85743

REBECCA K. WILSON, Herbarium, Department of Plant Sciences, University of Arizona, Tucson, Arizona 85721

DAVID A. YETMAN, Southwest Studies Center, University of Arizona, Tucson, Arizona 85721

INDEX

For infrequently cited genera and species, consult appendixes on pp. 72–99, 148–50, 168–69 (plants); 183–92 (herpetofauna); 192–97 (mammals); and 214–43 (birds).

beans (*Phaseolus* spp.), 155, 159, 163, 164
becards (*Pachyramphus* sp.), 208, 222
Begoniaceae, 74
Bignoniaceae, 71, 74, 127–28
birds, 10–11, 201–2; species status of, 213–14; variety of, 200–201, 202–3, 214–43
Bixaceae (Cochlospermaceae), 74, 128
black market: in herpetofauna, 182–83
Boidae, boa constrictor, 186, 190
Bombacaceae, 75, 128–29
Bombycillidae, 225
Boraginaceae, 75, 129
Bouteloua spp., 38, 43, 59, 63, 71, 95
Brahea aculeata, 42, 98, 143, 148
Brasica campestris, 154, 168–69
brasil (*Haematoxylum* spp.), 4, 6, 49, 53, 59, 65, 107, 149; human use of, 122, 123, 138–39
brea (*Cercidium praecox*), 107, 121, 124, 136–37, 148
Bromeliaceae, bromeliads, 5, 64–65, 71, 94
Brongniartia spp., 24, 25, 50 (fig.), 52, 55, 59, 65, 69, 82, 107, 136, 148
Buddlejaceae, 75
buffelgrass (*Pennisetum ciliare*), 3, 13, 14, 15, 60, 62, 63, 106, 135, 158, 211
Bufonidae, *Bufo* spp., 176, 180, 183–84, 189
Buichilame Josaino, Paulino, 113
Burseraceae, *Bursera* spp., 10, 71, 75, 148; distribution of, 23, 24, 25, 26, 48, 49 (fig.), 59, 69, 107, 108; human use of, 12, 116, 122, 124, 125, 129–31; seasonal changes in, 53, 55
Buteogallus spp., 206, 207, 213, 216, 230
Buteo spp., 206, 214, 216, 230–31
buttercup tree (*Cochlospermum vitifolium*), 53, 54, 74, 128

cacachila (*Karwinskia humboldtiana*), 55, 60, 107, 143, 149
caciques (*Cacicus* sp.), 206, 213, 229, 242
Cactaceae, cacti, 47, 49, 59, 61, 71, 75–76; columnar, 6, 11–12, 24, 48, 69, 131–32, 148. *See also various genera*

Caesalpinia spp., 12, 24, 53, 60–61, 69, 83, 107, 148; human use of, 113, 117, 122, 136
Callipepla douglasii, 209, 231
Calocitta colliei, 11, 176, 200, 210, 223, 238
Calypte costae, 203, 205, 220, 234
camaleones (*Phrynosoma* spp.), 181
Campanulaceae, 76
Campephilus spp., 205, 213, 221
Campylorhynchus spp., 206, 224, 235
candelilla (*Plumeria rubra*), 55, 74, 121, 127
Canidae, 196
Cannabinaceae, *Cannabis sativa*, 15–16, 66, 76
Capparaceae, 71, 76, 132
Caprimulgiformes, Caprimulgidae, 219, 234
Capsicum annuum, 61, 70, 91, 154 (table), 160–61, 168–69
Cardinalidae, 228, 241–42
Caricaceae, 76
Carnivora, 196–97
carving: wood for, 105, 130 (fig.), 131, 136, 139, 140, 143, 144, 145
Caryophyllaceae, 76
cascalosúchil (*Plumeria rubra*), 55, 74, 127, 149
Casimiroa edulis, 90, 108, 121, 144, 148
Cathartidae, 215, 230
Catharus aurantiirostris, 206–7
cattle, 8, 106, 115, 116
Caudata, 189
Ceiba acuminata, 4, 49 (fig.), 50 (fig.), 69, 75, 107, 148; human use of, 125, 128–29; seasonal changes in, 54–55
Celastraceae, 76, 132–33
Celtis spp., 43, 55, 59, 60, 93, 147, 148
Cercidium spp., 83, 107, 121, 124, 136–37, 148
Certhidae, 224
Cervidae, 197
chacalacas (*Ortalis* spp.), 10, 11, 209, 216, 231
Chamela, 21, 22, 23, 37, 47, 51, 69, 203; flora of, 66, 68, 70, 71; forest productivity of, 27–29
Chamela Bay, 26, 46

frogs, 175, 176, 177–78, 179, 180, 184, 189
frost, 5, 109
Fuerte, Río, 17, 38, 39
furniture: wood used for, 7, 13, 16–17 (figs.),
   116, 133, 135, 138, 139, 141, 145–46

Galliformes, 216, 231
garambullo *(Pisonia capitata)*, 43, 69, 116,
   143, 149
Gastrophryne olivacea, 179, 180, 184, 189
geckos, 180, 181, 185, 190
Gentianaceae, 81
Gentry, Howard Scott, 23, 37, 38, 47, 58
Geococcyx spp., 204, 218, 233
geology, 10, 23, 39, 43, 153
Geomyidae, 195
Geranospiza caerulescens, 209, 216, 230
Glaucidium spp., 205, 218–19, 233
gloria *(Tecoma stans)*, 121, 128, 150
gopher, Botta's pocket *(Thomomys bottae)*,
   173, 195
Gopherus agassizii, 176, 185, 189
Gouania rosei, 49, 61, 89, 97
gourds *(Lageneria* spp.), 154 (table), 160,
   169
grama *(Bouteloua* spp.), 38, 43, 59, 63
Gramineae (Poaceae), 70, 71, 95–98
granadilla *(Malpighia emarginata)*, 43, 107,
   142, 149
Gran Desierto, 65, 68, 69, 70
grasses, 43, 53, 58, 59, 60, 62–63, 64, 69;
   domesticated, 155–58
grassification, 8, 19. *See also* buffelgrass
grazing, 14–15, 107, 139, 211
grosbeaks, 208, 228, 229, 242
Gruiformes, 217
Guaiacum coulteri, 12, 48, 69, 107, 149;
   human use of, 121, 122, 123, 147–48;
   seasonal changes in, 53, 54
guamúchil *(Pithecellobium dulce)*, 9, 60,
   121, 141, 149
Guarijíos (Makurawe), 3, 11, 13, 137; crops
   of, 156, 157, 158, 159, 160, 161, 168–69
guásima *(Guazuma ulmifolia)*, 7, 10, 55, 59,

65, 69, 107, 149; human use of, 13, 116,
   123, 145–46
guayacán *(Guaiacum coulteri)*, 12, 48, 69,
   94, 107, 149; human use of, 121, 122, 123,
   147–48; seasonal changes in, 53, 54
guayavillo *(Acacia coulteri)*, 108, 135, 148
guayparín *(Diospyros sonorae)*, 60, 133–34,
   148
Guazuma spp., 7, 55, 59, 65, 69, 92, 107, 149;
   human use of, 116, 123, 145–46
Guerrero, 24, 30
güiloche *(Diphysa occidentalis)*, 13, 113,
   137–38, 148
güinolo *(Acacia cochliacantha)*, 60, 107, 135,
   148
Güirocoba, 42, 43, 45, 209, 211
gymnosperms, 73. *See also* sabinos

Habernaria quinqueseta, 41, 65, 98
habitats, 43; agricultural, 164–65, 166–
   67; aquatic, 57–58; of Sierra Madre
   Occidental, 152–53
hackberry. *See* cumbro
Haematoxylum spp., 6, 10, 49, 53, 59, 65, 84,
   107, 149; human use of, 122, 123, 138–38
Harpyhaliaetus solitarius, 207, 230
harvesting: of native resources, 7, 11–12
Havardia spp., 59, 84, 123, 139, 149
hawks, 206, 207, 209, 213, 214, 216, 230–31
Helianthus annuus, 154 (table), 168–69
Heliocarpus spp., 69, 93, 107, 123, 146–47,
   149
Heliomaster constantii, 205, 220, 234
Helodermatidae, *Heloderma* spp., 11, 177,
   178, 182, 185, 190
herbs: on Río Cuchujaqui, 41, 42, 43, 59,
   64–65, 69; understory, 49, 53–54
herons, 11, 202, 206, 214–15, 230
Herpailurus yagouarundi, 192, 196
herpetofauna, 173; conservation and, 181–
   83; myths about, 180–81. *See also by*
   family; type
Herpetotheres cahinnans, 209, 213, 216, 231
Heteromyidae, 196

hiedra *(Rhus radicans),* 10, 41, 74
Highway 16: gradsect study along, 21–22
Hintonia spp., 12, 25, 48, 50 (fig.), 53, 69, 90,
   107, 144, 149
Hippocrateaceae, 71
Hirundinidae, 223
honeybees, 121, 134, 144
Huastecans, 11
Huatabampo, 103, 106, 132
Huites Dam, 6
humidity, 56, 110
hummingbirds, 201, 202, 203, 205, 207,
   219–20, 234
hurricanes, 9, 43
hybridization: crop, 163–64
Hydrophyllaceae, 81–82
Hylidae, *Hyla arenicolor,* 178, 184, 189
Hypopachus variolosus, 177, 184, 189
Hyptis spp., 61, 69, 82, 154, 161, 169

Icteridae, 228–29, 242
Icterus spp., 208–9, 228–29, 242
igualama *(Vitex mollis),* 147, 150
Iguanidae, iguanas, 143, 176, 180, 182, 185,
   190
indigenous peoples, 3, 11–13, 102–4. *See also
   by culture; tribe*
Insectivora, 193
intercropping, 164–65
introduced species. *See* exotics
Ipomoea spp., 4, 5, 49, 59, 61, 62, 71, 79,
   107, 149; human use of, 121, 133; seasonal
   changes in, 53, 54, 66
Ismael, Hurricane, 9, 43, 146

jaboncillo *(Fouquieria macdougalii),* 53, 55,
   59, 81, 107, 122, 135, 149
Jacquinia macrocarpa, 52, 53, 59, 69, 92, 107,
   146, 149
jaguars *(Panthera onca),* 3, 11, 197
jaguarundi *(Herpailurus yagouarundi),* 192,
   196
Jalisco, 28, 46, 47, 66, 70, 157

Janos, 165
Jatropha spp., 24, 25, 43, 49, 50 (fig.), 59, 60,
   81, 107, 108, 149; human use of, 122, 134;
   seasonal changes in, 53, 55, 110
jays, 11, 223, 228. *See also* Magpie-Jay
Jesuits, 36, 102
Joba, 165
joso *(Conzattia* sp.; *Albizia* spp.), 10, 110, 136,
   148
Juncaceae, *Juncus tenuis,* 58, 63–64, 98

kapok. *See* pochote
Karwinskia humboldtiana, 55, 60, 89, 107,
   143, 149
Kinosternidae, *Kinosternon* spp., 179–80,
   185, 189

Labiatae, 71, 82
Lagenaria siceraria, 80, 154 (table), 169
Lagomorpha, 195
Lampropeltis spp., 187, 191
land races, 152–53; of maize, 155–58,
   163–64, 166
land use, 16–17, 103. *See also* agriculture;
   ranches, ranching
Laniidae, 222
La Ranchería, 43–44, 45, 59, 60, 61
Laridae, 217
Las Bebelamas, 106, 115–16, 145
Las Bocas, 104, 105, 113, 115
Las Lajitas, 42, 45, 57–58
Las Rastras, 105, 110, 112–13, 142
Leguminosae (Fabaceae), legumes, 8, 66,
   70, 71, 82–85; human use of, 12, 121–22,
   135–42; seasonal changes and, 53–54. *See
   also various genera*
Lemnaceae, *Lemna* spp., 57, 98
Lennoaceae, 85
Lepidocolaptes leucogaster, 205, 221, 235
Leporidae, 195
Leptodactylidae, *Leptodactylus melano-
   notus,* 179, 184, 189
Leptophis diplotropis, 143, 187, 191

ABOUT THE EDITORS

ROBERT H. ROBICHAUX is Associate Professor in the Department of Ecology and Evolutionary Biology at the University of Arizona. His research focuses on plant diversity, evolution, and conservation. He recently edited *Ecology of Sonoran Desert Plants and Plant Communities* (University of Arizona Press, 1999).

DAVID A. YETMAN is Associate Research Social Scientist in the Southwest Center at the University of Arizona. He specializes in studies of Sonora, Mexico, including rural development, ethnobotany, and ethnohistory. He recently edited *Gentry's Rio Mayo Plants: The Tropical Deciduous Forest and Environs of Northwest Mexico* (University of Arizona Press, 1998) and wrote *Scattered Round Stones: A Mayo Village in Sonora, Mexico* (University of New Mexico Press) and the forthcoming *Trails of the Makurawe: The Guarijíos of Sonora and Chihuahua* (University of New Mexico Press).